Risks, Dangers, and Rewards
in the Nova Scotia Offshore Fishery

According to Labour Canada, workers in the offshore fishery are more likely to be injured than workers in mining, construction, or forestry. Yet until recently these casualties at sea have been largely ignored by government and labour organizations. *Risks, Dangers, and Rewards in the Nova Scotia Offshore Fishery* describes the hidden cost paid by workers in the Nova Scotia offshore fishery, a cost measured not in dollars and cents but in deaths and injuries.

In this comprehensive study Marian Binkley documents the level of risk and assesses the general health and stress level of workers in the Nova Scotia offshore fishery. She considers both direct and indirect factors, including shipboard working environment; stress; accidents, injuries, and general health; awareness of safety; job satisfaction and family life; and the impact on working conditions of government resource policies and companies' scientific management strategies.

Using statistical analysis, participant observation, surveys, and both formal and informal interviews, Binkley establishes that factors such as home and community life and the unforeseen consequences of changes in technology and management of the fishery affect the immediate work experience of fishers and can increase the dangers of an already hazardous occupation.

MARIAN BINKLEY is professor of social anthropology, Dalhousie University.

Risks, Dangers, and Rewards in the Nova Scotia Offshore Fishery

MARIAN BINKLEY

McGill-Queen's University Press
Montreal & Kingston • London • Buffalo

© McGill-Queen's University Press 1995
ISBN 0-7735-1313-2

Legal deposit fourth quarter 1995
Bibliothèque nationale du Québec

Printed in Canada on acid-free paper

This book has been published with the help of a grant
from the Social Science Federation of Canada, using
funds provided by the Social Sciences and Humanities
Research Council of Canada.

McGill-Queen's University Press is grateful to the
Canada Council for supporting its publishing program.

Canadian Cataloguing in Publication Data

Binkley, Marian, 1950–
  Risks, dangers, and rewards in the Nova Scotia
offshore fishery
  Includes bibliographical references and index.
  ISBN 0-7735-1313-2
  1. Fishers – Health and hygiene – Nova Scotia.
  2. Fishers – Wounds and injuries – Nova Scotia.
  3. Fishers – Job stress – Nova Scotia. 4. Fisheries –
  Accidents – Nova Scotia. I. Title.
  SH224.N8B54 1995 363.11'96392'09716 C95-900372-X

Typeset in New Baskerville 10/12
by Caractéra production graphique, Quebec City

To the men of the *Cape Aspey*

Contents

Figures and Tables

Acknowledgments

Risks, Dangers, and Rewards was a team effort. Although the membership of the team changed through time, each member left his or her stamp on the project. Jennifer Dingle, the first project manager, supervised the 1986 survey, which was administered by fourteen trained interviewers hired from Statistics Canada, and she set up the computer database derived from that survey. David Millar, the second project manager, conducted the 1987 survey and the in-depth interviews with Fred Winsor and myself and added the 1987 data to the database. Shawna Burgess held the fort as our trusty receptionist/ secretary during the data collection phase. Annette Cossar, Marian MacKinnon, Mary Morash, and Cathy Smith transcribed and deciphered the interview tapes. Final manuscript preparation was made easier by the assistance and advice of Susan Brown and Donna Edwards at Dalhousie, Peter Blaney and Joan McGilvray at McGill-Queen's University Press, and Avivah Wargon.

The project was supported by grants from the Social Sciences and Humanities Research Council of Canada and the Dalhousie University Research and Development Fund.

During this project a number of colleagues gave me encouragement and advice: Richard Apostle, Fay Cohen, Jack Crowley, Ann Dwire, Leonard Kasdan, Barbara Neis, Victor Thiessen, and the three anonymous manuscript reviewers. Personnel from a number of government agencies supported this project: the federal Department of Fisheries and Oceans; Marine Casualty Investigations Division; the Nova Scotia Ministries of Fisheries and Labour; and the Nova Scotia Workers' Compensation Board. Without the cooperation and help of managers of the major fish companies and members of fishermen's unions this project could not have been possible. Finally, *Risks, Dangers, and Rewards* is about deep sea fishers in Nova Scotia, and it could not have been done without their cooperation and help and that of their wives and families.

Risks, Dangers, and Rewards
in the Nova Scotia Offshore Fishery

CHAPTER ONE

Introduction

Risks, Dangers, and Rewards examines the working conditions of Nova Scotia deep sea fishers and the high price they pay for employment: a price measured not in dollars and cents, but in deaths and injuries sustained by workers in the industry. When I began this research in 1986, 14,859 licensed fishers laboured in the harvesting sector,[1] and about half (7,734) of these men worked full time in the deep sea sector (Canada, Department of Fisheries and Oceans, 1991, 111–13). In 1987, the Atlantic Canada fishery, one of the world's most productive, accounted for approximately 3 percent of the world's total catch, 80 percent of Canada's total catch, and two-thirds of Canada's total fish exports (Northwest Atlantic Fisheries Organization 1989; United Nations, Food and Agricultural Organization 1987). Traditionally the fishery employs, directly and indirectly, the largest number of workers in Nova Scotia, a region of high unemployment. Fishing combined with work in onshore processing plants and employment in indirectly related enterprises forms the economic and social backbone of the small communities nestled along Nova Scotia's indented coastline. The rise and fall of the fortunes of the fishery dictates the economic prosperity or decline of the coastal communities dependent on it.

Nova Scotia deep-sea fishers comprise a separate occupational community. Pilcher (1972) argued that workers' self-identification, their commitment to their occupation, and its importance in their lives define separate occupational groups. Nova Scotia deep sea fishers identify themselves very specifically as "offshore fishermen" and refer to their fellow crew members as their "other" family. Although they come from many different parts of Nova Scotia, they spend a long time away from their families, friends, and communities, working together in an isolated and dangerous environment. They share a bond with other men who work on deep sea fishing vessels, with whom they have working conditions and problems in common. But they distance themselves from coastal fishers, whose work deep sea fishers see to be neither as difficult nor as challenging as theirs (frequently referring to the coastal fishery in derogatory terms, e.g., the "mosquito fleet"). The deep sea fishers' distinctive clothing – floater vests, hard hats, steel-toed shoes, foul-weather gear – also separates them from other occupations. Deep sea fishers present a united front to outsiders and share a common identity, although they perform many different types of work within their community.

Fishing and land-based occupations differ in another important way. Few workers in land-based occupations daily confront comparable working conditions, especially the possibility of losing their lives. At sea the working environment cannot be controlled and the vessel always makes complicated motions, even on the calmest water. Fishers work in close proximity to heavy fishing gear, equipment, and machinery, either on an open deck awash with cold salt water or below deck in the cramped damp fishroom or in the refrigerated hold. They accept the constant risk of being swept overboard, of being thrown into the machinery, of slipping and falling, and of having gear, cables, and machinery fall on top of them. Even the most advanced fishing vessel can be lost at sea.

Moreover, the risks associated with fishing exceed those of other seafaring occupations. Warner (1983, 73) describes the risks in the North Atlantic deep sea fishery by noting that "The kind of work which fishermen routinely perform in bad weather has no parallel in the world of seafaring. On a merchant ship such feats as leaping to catch wild cables or ducking out of the way of objects shifting around on deck are normally required only in emergencies, but aboard trawlers they form part of nearly every shooting and hauling of nets."

But just how dangerous is it? Statistics indicate that it is one of the most hazardous and physically demanding occupations in Canada. Each year more workers in the deep sea fishery die or sustain injuries from work-related accidents than in any other industry. Until recently these casualties were largely ignored by government and labour when legislating safety regulations and compensation. According to Labour Canada (1983, 57), in 1981 147.1 per 100,000 deep sea fishers died from industrial accidents compared with 70.8 per 100,000 miners, 36.6 per 100,000 construction workers and 91.5 per 100,000 forestry workers. In 1985 212 per 100,000 deep sea fishers died, compared with 74 per 100,000 miners, 32 per 100,000 construction workers and 118 per 100,000 forestry workers (Yoon Kim, Labour Canada, personal telephone communications, 1987). The Canadian experience is not unique. Thompson and co-workers (1983) reported that British fishers had a fatality rate twenty times higher than that of workers in the manufacturing industries. In the United States the annual fatality rate for commercial fishers approximates 80–85 per 100,000, about seven times higher than the average for all industries (MacDonald and Powers, 1990, 1).

RESEARCH METHODS

In order to generate an integrated picture of the research problem, this project employed four complementary methods: statistical analysis, participant observation, surveys, and formal and informal interviews (see figure 1.1). I generated preliminary demographic and socio-economic profiles of deep sea fishing crews, physical characteristics of their vessels, and the basic parameters of their catches from data collected from the federal Department of Fisheries and Oceans (DFO) and from the Nova Scotia Department of Fisheries. Information on the type and frequency of injuries, accidents, and deaths related to deep sea fishing came from the Marine Casualty Investigations Division of the Canadian Coast Guard (Department of Transport) and the Nova Scotia Workers' Compensation Board. This information, combined with data from a preliminary ethnographic study and informal interviews with workers, formed the basis for the subsequent survey, questionnaires, and interview schedules, and supplied valuable background information on the population.

I conducted informal interviews with officials from the federal Department of Fisheries and Oceans, the Canadian Coast Guard, the Nova Scotia Department of Fisheries, and managers in fish companies. In order to formulate hypotheses to be tested during the project, I added this information to the statistical analysis of the government material mentioned above. Throughout the pretesting of the questionnaire, I gathered information from participant observation to help focus and restructure the interview schedule. Once interviewing had begun, I used participant observation to assess the reliability of responses recorded during the interviews and to check on apparent inconsistencies between actual and reported behaviour.

I developed a precoded general survey of fifty-nine items covering the following topics: demographic characteristics and work experience of the crews; organization of work, general working conditions, and job satisfaction; attitudes toward safety and training in safety procedures and equipment usage; types of accidents, injuries, and deaths that occur on vessels and the circumstances of these incidents; general

Statistical sources	federal	Marine Transport
		Department of Fisheries and Oceans (DFO)
	provincial (Nova Scotia)	Department of Fisheries
		Workers' Compensation Board (WCB)
Participation and observation		ethnographic material about fishing vessels and deep sea fishing communities
Surveys	1986	general survey of deep sea fishers ($N=334$)
	1987	leavers survey ($N=92$)
Interviews	formal	well workers ($N=50$)
		leavers ($N=92$)
	informal	provincial and federal fisheries officers
		federal DFO observers
		company managers
		medical personnel
		union officials
		fishers' wives

Figure 1.1
Research methods used in project

health of the crew; social and economic relationships between the crews and other sectors of the industry; and community and family responsibilities and experiences of crew members.[2]

A stratified random sample drawn from all fishing vessels over 45 feet registered in Nova Scotia and licensed as deep sea vessels produced six major groups of vessels according to type of gear and vessel size: groundfish trawlers over 65 feet, groundfish trawlers under 65 feet, groundfish and scallop draggers under 65 feet, scallop draggers under 65 feet, scallop draggers over 65 feet, and herring seiners. I chose a representative sample of vessels from each group, reflecting geographical distribution as well as vessel size, gear type, and crew size.

In January and February of 1986 I identified these vessels through the Registry of Joint Stock Companies, the Nova Scotia Department of Fisheries' representatives and contact with owners. The sample excluded those vessels that had been condemned, those that did not fish out of Nova Scotia ports, and those that could not be traced. I identified the captains and one-third of the crew of each selected vessel, and, where possible, I selected crew on each vessel who represented a variety of job designations. These designations corresponded to three categories: captains/mates, including substitute mates and captains; other officers, including first and second engineers, bosuns, cooks, and substitutes; and crew, including trawlermen, deckhands, icers, winchmen, and learners. Because of the concentration of deep sea work in southwest Nova Scotia, the sample focuses on this area.

Fourteen trained interviewers who lived in or near the various fishing communities conducted the interviews during March and May of 1986. All had experience with similar questionnaires, and Statistics Canada had previously employed many of them for both census and labour-force studies data collection.[3]

The data collected had inherent biases. First, another study on fishers taking place in the Digby area cut across the sample. Many of the people the interviewers contacted had been asked to take part in the other survey, sometimes even by the same interviewer. This situation lowered the response rates. Since many of the Digby fishers worked on herring seiners, this resulted in an underrepresentation of herring fishers in the survey, so I excluded them from the sample. Second, a number of individuals the interviewers contacted no longer fished, so I excluded them from the analysis as well. Interviews with deep sea fishers had to take place during the time they had on shore, usually 48 to 72 hours, and that meant a diminished response rate. (This difficulty had also been noted by Poggie and Pollnac [1978], Apostle, Kasdan, and Hanson [1985], and Binkley [1989].) I based the analysis on 334 completed interviews from a stratified random sample of 480 possible respondents. The overall completion rate of the survey reached 70 percent, compared to 87.4 percent in Apostle, Kasdan, and Hanson (1985, 257).

Table 1.1 indicates the basic demographic characteristics of the sample broken down by fishery, job designation, and enterprise size. All fishers worked full-time and spoke English. As table 1.2 shows, 84 percent were married (including common law), 11 percent were single, and 5 percent were separated, divorced, or widowed. This may be an artificially low number of separated or divorced fishers, and may be attributable to the interviewing procedure.[4]

The mean age of the sample group was 33 years. Over 57 percent of the workers were in their twenties and thirties, slightly less than 30 percent were in their forties, and approximately 12 percent were fifty and over. Only two workers were under twenty (about 1 percent). The mean level of education was just under nine years of schooling; about 52 percent had grade 8 or less, 8 percent had finished high school and a little over 1 percent had gone to college or university.

Men working on vessels of less than 65 feet were slightly younger (32.3 years) than men working on the larger scallopers (33.7) and

Table 1.1

Basic demographic characteristics for sample of Nova Scotia deep sea fishers

Table 1.1a

Mean level of education and age by fishery

Fishery	Education[1]		Age[1]	
	Mean	(N)	Mean	(N)
Trawler over 65 ft	8.2	117	32.3	118
Scalloper over 65 ft	8.1	153	33.7	154
Midshore	9.7	53	32.3	59
Total population	8.3	328	32.8	331
	Missing 6		Missing 3	
	$F = 10.1$		$F = .6$	
	$p = .000$		$p = .542$	

Table 1.1b

Mean level of education and age by job designation

Job designation	Education[1]		Age[1]	
	Mean	(N)	Mean	(N)
Captains and officers	8.7	148	36.1	148
Crew	8.2	183	30.3	185
Total population	8.4	331	37.1	333
	Missing 3		Missing 1	
	$F = 6.3$		$F = 23.9$	
	$p = .013$		$p = .000$	

Table 1.1c

Mean level of education and age by enterprise size

Enterprise size	Education[1]		Age[1]	
	Mean	(N)	Mean	(N)
Large enterprise	8.2	262	33.5	262
Small enterprise	9.5	67	30.3	67
Total population	8.4	331	32.8	331
	Missing 3		Missing 3	
	$F = 21.7$		$F = 4.4$	
	$p = .000$		$p = .037$	

Table 1.1
(continued)

Table 1.1d
Distribution of family income by job designation, vessel type, and enterprise size

	Job designation[2]		Vessel type[2]			Enterprise size[2]	
Family income	Captains and officers	Deckhands	Trawler +65 ft	Scalloper +65 ft	Midshore	Large	Small
Number	(149)	(181)	(117)	(152)	(58)	(261)	(67)
Less than $10,000							
N	2	10	5	6	0	9	2
%	1.3	5.5	4.3	3.9	0	3.4	3.0
$10,000–$24,999							
N	27	103	31	78	21	103	27
%	18.1	56.9	26.5	51.3	36.2	39.5	40.2
$25,000–49,999							
N	76	64	58	53	27	111	28
%	51.0	35.4	49.6	34.9	46.6	42.5	41.8
$50,000–74,999							
N	31	3	16	10	8	26	8
%	20.8	1.7	13.7	6.6	13.8	10.0	12.0
$75,000 or more							
N	13	1	7	5	2	12	2
%	8.7	0.6	6.0	3.3	3.4	4.6	3.0

Note: Total number of cases per variable shown in brackets.

$\chi^2 = 81.8$ $p = .000$ $\chi^2 = 21.8$ $p = .005$ $\chi^2 = .6$ $p = .996$
$DF = 4$ $DF = 4$ $DF = 4$

[1] Variance measured by F-test.   [2] Variance measured by chi-square test.

trawlers (32.3). These men were also better educated, with an average of 9.7 years of schooling versus 8.1 and 8.2 respectively. Captains and officers had statistical significantly better education than crew members. Captains, and officers, were significantly older than crew. Those individuals working for small enterprises were younger and significantly better educated than those working for larger enterprises.

I found no significant difference in income distribution between those who worked for small or large enterprises, although workers on trawlers and midshore vessels usually earn significantly more than workers on scallopers. As illustrated in table 1.1(d), the median income for captains and officers fell in the $25,000 to $49,999 range, while the median income for crew members was between $10,000 and $24,999. Approximately 27 percent of captains had incomes over $50,000 and 12 percent had incomes over $75,000. Twelve percent of officers had incomes between $50,000 and $74,999. A little over

31 percent of crew members had incomes over $25,000.[5] For over 81 percent of those interviewed, their income from fishing comprised the total family income. Less than 10 percent of the workers reported that more than a quarter of their family income came from sources other than their fishing (i.e., other family members' income or their own income from other sources).

In 1987 we interviewed workers who had left the deep sea fishery for health, economic, or other reasons. Fieldwork took place between May and October of 1987. Two trained male interviewers worked as a team out of two deep sea fishing communities – one in Canso, the other in southwest Nova Scotia. One of these men had some experience in interviewing and had worked in the deep sea fishery both as a deckhand and as a Department of Fisheries and Oceans officer. The other had extensive interviewing experience and had spent some time on midshore and coastal vessels.

The interviewers made two separate visits: one to administer a survey, and another to conduct an in-depth formal interview. They interviewed workers who had left the fishery during the last five years. The topics covered in the 1987 survey, forty-six items in all, included relevant demographic characteristics, past work experience, job satisfaction, working conditions, specifics of injuries and treatment (where applicable), medical histories, past and present family responsibilities, support networks, and employment and retraining opportunities. Throughout the interviews, they paid special attention to the sensitive nature of the material being solicited, and they accommodated the workers' values and sense of dignity, and their concerns about insurance and employment benefits. With the consent of the respondents, the fieldworkers taped all of the interviews. They then entered the interview summaries and the survey in computer-readable form for analysis.

We gathered information using the snowball method,[6] a technique used increasingly in community studies that are defined occupationally as well as geographically. Although this method is often criticized as random, the use of quotas ensures some representativeness.[7] I identified our initial contacts from information given to us by company and union officials and this group provided additional names. I wanted a sample of one hundred people with a crew-to-officer ratio of three to one, in order to reflect the typical ratio found on deep sea fishing vessels.

They interviewed 101 workers, but I reduced the sample to 92 because of incomplete data. All of the sample had worked full time for at least one year in the deep sea fishery, and had left in the last five years. Of these fishers, three out of four had left because of an

Table 1.2
Comparison of demographic characteristics of deep sea fishers by turnover status

Demographic characteristics	Current workers	Former workers
Mean age	33	40
Age distribution		
% under 40	60	25
% between 40 and 50	30	25
% 50 and over	10	50
Marital status (%)		
Married	84	76
Single	11	10
Divorced/separated	5	14
Education (years)	9	8

accident or injury, and the remainder had left for other reasons, ranging from illness to economic incentives in the other fisheries. While working the deep sea, 30 percent had been officers or captains, and 70 percent had been crew members. The majority of crew members (73%) had left because of injuries, while captains and officers had left mainly for economic reasons.

Of the workers interviewed 76 percent were married (including common law), 10 percent were single and 14 percent were divorced, separated or widowed. The average captain/officer was forty-seven years old, whereas the average crew member was a decade younger. The level of schooling averaged just under eight years for both groups.

Table 1.2 summarizes selected demographic characteristics of the two samples of the 1986 and 1987 surveys respectively. The mean age and the age distribution of those samples varies substantially and reflects the nature of deep sea fishing, which is characterised as a young man's fishery. Most deep sea fishers say they started fishing as soon as they left school. As one informant put it, "If you aren't out by forty-five, you should be."

During the project, I conducted informal interviews with informants from six distinct groups: federal and provincial Department of Fisheries fieldworkers; federal Department of Fisheries and Oceans observers; fish company managers and other officials; medical personnel (e.g., physicians, public health nurses) working in the areas where the fishing vessels and the crews originate; union officials; and fishers' wives. DFO observers and fieldworkers from the provincial and federal Departments of Fisheries described types of gear, work organization, government regulations, and personal experiences. The former group sail on many vessels and have a wide range of experience with different types of fisheries; the latter work in fishing communities and in all the fisheries. I compared their information with the individual case histories of workers. Medical personnel supplied information on the general health of deep sea workers and the kind of accidents, illnesses, and deaths associated with work on those vessels. I also asked them about the general health of the local people and the effects, if any, that fisheries-related health problems have on the community as a whole.

I handled all the information collected in such a way as to guarantee privacy. Interviewers obtained informed consent by explaining the nature of the study in detail before collecting any interview or survey information. None of the names and addresses of respondents appear on the interview schedules or questionnaires, nor do they appear in any stored data. In order to disguise identities, I changed all the interviewees' names and slightly altered the details of their personal lives and histories. When each stage of the study had been completed, I sent a summary of the findings of that stage to each of the respondents who requested it.

Throughout the text I have quoted material from the interviews. At the end of each quote I have cited the source of the interview by the groupings described in figure 1.1 (e.g., well workers, leavers, fishers' wives). In the case of fishers, I have also included their job designation (e.g., captain, mate, deckhand).

DESCRIPTIVE OUTLINE OF TEXT

Research into risk in the North Atlantic fishery has focused on three areas: technological change and its effect on working conditions (Andro et al. 1984; Neis 1987; Poggie 1980); the interaction of socioeconomic factors with working conditions, in particular job satisfaction (Apostle, Kasdan, and Hanson 1985; Binkley 1990b; Gatewood and McCay 1988, 1990; Pollnac and Poggie 1988b); and social and economic factors influencing fishers' awareness and perception of physical risks (Poggie and Gersuny 1972; Pollnac and Poggie 1988a). Smith (1988) attempted to expand and to connect these topics by redefining the concept of risk in the fishery in socio-economic as well as physical terms.

Risks, Dangers and Rewards divides into two parts: (1) discussion of background information and the work environment, and (2) discussion of health and safety issues and concerns.

Chapter 2 discusses the theoretical and historical background for the study.

Chapter 3 focuses on the work environment and technology with special attention to variations in the organization of work and in gear and vessel types, the job designation of workers, and the amount of capitalization in the enterprise. It also includes a discussion on the impact of government resource policies and scientific management strategies on working conditions.

Chapter 4 explores the impact of fishers' work schedules on their families, where the normal work schedule usually consists of ten days at sea followed by forty-eight hours on land.

Chapter 5 continues the discussion of working conditions, using survey data to present a summary of workers' satisfaction with present working conditions. It also explores the job satisfaction of workers who have left the deep sea fishery.

Chapter 6 uses ethnographic data to examine the specific trade-offs fishers make between monetary and nonmonetary benefits and how the work environment limits these trade-offs. It concludes by discussing some of the underlying reasons why well workers leave the deep sea fishery.

Chapter 7 discusses the inherent biases of federal and provincial government statistics on working conditions, accidents, and casualties in the Nova Scotia fishery in general and the deep sea fishery in particular. The discussion examines the prevailing attitudes in various segments of the industry and in governmental agencies toward health and safety concerns, and concludes by describing the types of injuries fishers experience and their relationship to work patterns.

While chapter 7 provides documentation of the actual level of risk within the fishery, chapter 8 explores the general health and the stress levels of deep sea fishers. It begins by examining general health patterns, sources of physical and emotional stress, and high-risk health behaviour such as drinking and smoking. It also discusses fishers' perceptions of risks and their awareness of safety concerns in the fishery, as well as whether low levels of awareness can be psychologically adaptive for fishers.

Chapter 9 focuses on fishers' awareness of safety concerns, how that awareness can be enhanced, and family members' concerns and perceptions of safety. The concluding chapter discusses the possible future of the deep sea fishery.

CHAPTER TWO

Theoretical Considerations

Industrial sociology, especially work on labour process following the work by Braverman (1975), focuses mainly on the male industrial workforce in manufacturing and primary industries (e.g., Burawoy 1972; Clement 1981). The general argument in these studies focuses on the labour process per se, and not on other factors relevant to studying the work experience such as marital status, parental status, and ethnicity (Zimbalist 1979; Burawoy 1985). Feminists and others have sharply criticized this restricted viewpoint, arguing that both the theory of the division of labour (Barrett 1980; Beechey 1987) and empirical research (Armstrong and Armstrong 1978; Connolly 1978; Lacasse 1970) make it clear that studies must include both analysis of the work process and the impact of other external factors.

The labour process necessarily forms the core of any discussion of occupational health and safety. But focusing on this alone limits understanding of the issues involved. The workplace is made up of the physical environment, the organizational environment, and the socio-economic environment. Individuals working in the deep sea fishery are also involved with their communities and their families. All these factors must be taken into account when discussing occupational health and safety issues.

In order to understand occupational illnesses and injuries one must examine the social relationships that surround occupational health and safety. The reproduction of labour depends on the physical maintenance of workers, including their health, ability to work, and job satisfaction (Doyal 1979). Any discussion of occupational illnesses and injuries must address the labour process – specifically, the relationship between capital's control of labour and the labour process, the state's support for and challenges to that control, and labour's response to both.

Doyal (1979, 39) argues that "acceptable" or adequate levels of reproduction of labour power will vary according to the supply of labour, the type of technology in use, the cost of illness, and workers' expectations of the level of health they should enjoy. In a similar vein, Schatzkin (1978) claims that levels of workers' health and governments' expenditures on medical care vary in relation to levels of unemployment and other circumstances influencing the ease with which workers can be replaced. Both government support for "Medicare" and the local level of care available to workers directly influence the costs to employers. The level of workers' health also depends on their capacity to organize (in workers' groups or unions), to struggle against unsafe and unhealthy work situations, and to demand better working conditions.

Employers have a vested interest in workers' health. As Doyal (1979) argues, it is not in the interest of capital to ignore the hazards of the workplace, since the creation of surplus value is dependent on workers' labour with its investments in skill and experience. However, any expenditures that promote the health and safety of workers must be financed from surplus value. Employers must "balance the costs and benefits of promoting health against the costs of productivity losses as a result of illnesses" (Walters 1985, 74).

Employers must also take into account the management strategies of their competitors: "Individual capitalists could jeopardize their competitive position unless all employers were compelled to meet common standards. In consequence, employers generally resist the recognition of hazards and introduce the least costly method of control" (Walters 1985, 59). Thus it is usually governments or unions, or both, that attempt to compel employers to make improvements. Employers resent and resist such intrusion. They seek to retain control over labour (Murray 1982) and to determine the nature and timing of any improvement in the labour process (Walters 1985, 59). Worker groups and unions must balance the hazards of the workplace against the risk of losing employment. If employers deem labour's demands for health and safety too costly, they will be compelled to seek an alternative labour supply.

One of the state's roles is to mediate between labour and capital in such conflicts. As a functional response to a capitalistic economy, the state must weigh workers' health needs against the accumulation of capital needed to maintain and stimulate the economy. The development of social policy plays out the class struggle. In response to pressure from labour, the state grants concessions in the form of social policy in an effort to restore social harmony (Gough 1979; Navarro 1978; Piven and Cloward 1971; Swartz 1977; Walters 1983).

THE PRESENT SITUATION

The Nova Scotia deep sea fishery demonstrates all of those social relationships. Until recently governments, employers, and labour have neglected to include fishers' occupational health and safety concerns in the debate over improved health and safety in the workplace. There are several apparent reasons for the exclusion of those concerns from legislation: jurisdictional wrangling between provincial and federal governments and within federal ministries; the inability, because of their mandates, of the Nova Scotia Workers' Compensation Board and the federal Department of Transport to address fishers' occupational health and safety concerns directly; the inability of the Workers' Compensation Board to define fishers as workers; and the inability, until quite recently, of fishers to form unions.

The fishery is under federal jurisdiction, and until recently the Department of Transport – not the Department of Labour – represented fishers' concerns. Legislation initiated by Labour Canada regarding safety and work practices applied only to industries on shore, and the Department of Transport did not develop parallel legislation for marine workers. With the development of the offshore oil industry in the 1980s and the subsequent *Ocean Ranger* disaster, the federal and provincial governments started paying serious attention to the hazards of working at sea. At the same time, companies and workers in the fishery began to address the problem of legislative jurisdiction over the health and safety concerns of deep sea workers. Revisions in 1987 of Part IV of the *Canada Labour Code* attempted to deal with this problem.[1]

Until 1987 no government agency had the specific task of collecting information on how many men were killed or injured in the fishery. Two state agencies do collect and report indirectly on the injuries and fatalities of Nova Scotia deep sea fishers – Marine Casualty Investigations, a division of the Canadian Coast Guard and part of the federal Department of Transport, and the Nova Scotia Workers' Compensation Board. Marine Casualty Investigations examines and records all incidents as "marine casualties" (accidents to vessels) and "accidents aboard ship" (accidents to people). The Nova Scotia Workers' Compensation Board provides an insurance scheme to compensate workers for job-related injuries and illnesses. However, these primary mandates, which do not include collecting statistics on injuries and deaths in the fishery, influence their working definitions of what constitute job-related accidents, injuries, and illnesses in the industry. Those definitions are also influenced by the agencies' role as mediators. Collecting data and reporting on them are by-products of fulfilling their primary functions. When marine occupations came under Part IV of the *Canada Labour Code*, Labour Canada became responsible for collecting statistics on work-related accidents and injuries.

Another and equally important reason for the exclusion of fishers from the debate on workers' health and safety lies in their traditional "co-adventurer" status. Fishers are paid on a share system. Each fisher, or "shareman," receives a specific portion or "share" of the proceeds of the catch after expenses have been deducted. Proponents of the "co-adventurer" position argue that fishers work not as employees but for themselves, taking the same risks and sharing in the same profits as vessel owners. Fishers contribute their labour, while the vessel owner provides the capital for the venture. This traditional "non-employee" status gained legal recognition in the 1947 revision of the *Trade Union Act*, which excluded fishers from participating in the Nova Scotia workmen's compensation programs and from forming unions. In 1971, with further revisions to the *Trade Union Act*, those restrictions disappeared (Nova Scotia, *Trade Union Act*, 1972, s. 1(*k*)(i)).

The fishery still presents particular problems for Workers' Compensation Boards, and they continue to characterize many fishers as self-employed and not as workers. Under a strict definition of the Act, a "shareman" or an independent operator is not a worker. Each individual provincial act has attempted to redefine the term "worker" so that at least some fishers will be covered. In Nova Scotia, fishers were first included in 1968 (Nova Scotia, *Workmen's Compensation Act*, 1968) and the following definition is currently in place:

(w) "worker" ... in respect of the industry of fishing, includes a person who becomes a member of the crew of a vessel under an agreement to prosecute a fishing voyage in the capacity of a sharesman or is described in the shipping articles as a sharesman or agrees to accept in payment for his services a share or portion of the proceeds or profits of the venture, with or without other remuneration, or is employed on a boat or vessel provided by the employer (Nova Scotia, *Workers' Compensation Act*, R.S.N.S. 1989, s. 2).

Even so, the current Nova Scotia Workers' Compensation Board legislation still does not cover many fishers because they work on small boats with crews of less than three or four, "workplaces" too small to be covered. In practice, this limitation excludes coastal fishers.

But the major reasons for fishers' current exclusion come from the past social relations within the industry. They are the historical product of governments' mediation between the interests of the fish companies and those of the fishers. In order to understand those relationships, we must look at their historical development.

THE HISTORICAL BACKGROUND

Before World War I the Nova Scotia fisheries, like all of the Atlantic Canada[2] fisheries, consisted primarily of a coastal small boat fishery, a small side trawler fleet, and a schooner fleet. In line with the recommendations of the 1920s Royal Commission to Investigate the Fisheries of the Maritimes and the Magdalen Islands, the federal government of the day severely restricted the development of the domestic deep sea trawler fleet. Following World War II, government policy changed dramatically to favour an industrialized and techno-logically advanced trawler fleet. This change made way for the rapid expansion of a deep sea fishing fleet, but it also signalled the begin-ning of the decline of the coastal fishing fleet and the demise of the obsolete schooner fleet.

The first *Workmen's Compensation Act* of Nova Scotia, enacted in 1917, was revised two years later to include fishers working on schooners and trawlers. Before this, either the federally sponsored Sick Mariners' Fund or the Lunenburg Fishermen's Relief Society insured Nova Scotia deep sea fishers. Both plans had low participation rates and minimal benefits (Winsor 1987). The Workmen's Compensation Board's plan represented a substantial improvement for fishers.

Early reports of the Workmen's Compensation Board show the perilous nature of the fishery. In 1920 two trawlers with a total of thirty persons on board sank. From 1921 to 1925 thirty-three Nova Scotians working on fishing vessels drowned. The August gales in 1926 claimed two schooners and fifty men. In 1927 four schooners sank and eighty-eight persons drowned in a single storm. Although all vessels sailed from the Lunenburg area, not all fishers hailed from there, and the board only paid compensation for a fisher's death if his family lived in Nova Scotia (Winsor 1987).

In the aftermath of the 1920 sinkings, the board reduced the amount of compensation paid to fishers from the levels set in 1919. After the 1926 disasters, the board attempted to double the rate of the premium paid by the fishing companies from $5 to $10 per $100 of wages paid. The companies argued that they could not pay this substantial rate increase and still be financially sound. As a result the provincial government set up the Royal Commission on Ratings of the Lunenburg Fishing Fleet and the Lumber Industry to look into the rates paid by both the Lunenburg fleet and the lumber industry (which was experiencing similar difficulties). Testimony given before Commissioner Dennis describes the fishery as an industry in economic trouble. The owners and captains argued that given the current market rates for fish and the disasters of the previous year, it was impossible for them to pay the new rates. But their pleas failed. Citing fiscal responsibility, the government granted the companies a one-year extension of the 1926 rates ($5 per $100 of wages paid), but set the rates for 1927 at $11 per $100 of wages paid and for 1928 at $20 per $100 of wages paid.

The owners lobbied the government extensively. As a result of this pressure the Board excluded fishers from the main body (Part I) of the *Workmen's Compensation Act* and developed a new section of the Act (Part III). Part III set out the terms and conditions of private insurance plans for fishers.[3] The benefits under the new plan provided substantially less support than those benefits that had been available under Part I of the Act, even at the reduced 1921 levels. Deep sea fishers did not receive full benefits under Part I of the Nova Scotia *Workmen's Compensation Act* until 1972. In the interim they had to rely on private insurance plans as set out in Part III.

During the debates over benefits paid by the board, rates paid to the board, and the inclusion/exclusion of fishers from Part I, no fishers took part – nor did they form any unions or worker groups before or during this time. Only owners and employers lobbied the government for these changes.

As the provincial royal commission conducted its inquiry, the federal Royal Commission to Investigate the Fisheries of the Maritimes and the Magdalen Islands was also examining the eastern fisheries. Since fisheries come under federal jurisdiction, some suggest that the Nova Scotia government had hoped that the federal government would take on the responsibility for fishers' health, safety, and compensation. Although Premier Rhodes alluded to this possibility in the 1928 Speech from the Throne, the federal commission stated in its report that fishers' health and safety lay outside its mandate (Canada, Royal Commission to Investigate the Fisheries 1928).

Subsequent reviews of the Workmen's Compensation Board as it pertained to fishers' concerns fared no better. In 1937 the Royal Commission on Workermen's Compensation looked into the nature of the Workmen's Compensation Board and made no mention of the fishery. In 1958, the Royal Commission to Inquire into the Workmen's Compensation Act of Nova Scotia received briefs from the fishing industry. Those presentations again put forward the views of the employers and owners of the fishing vessels and the insurance companies managing the funds under Part III. They cited the high cost of workmen's compensation payments and the smooth workings of the private plans as evidence for continuing the inclusion of fishers under Part III. No submissions by or on behalf of fishers were filed.

The absence of worker groups or unions is a crucial element in the history of workers' health and safety legislation in this industry. In 1946 and 1947 the Canadian Fishermen's Union attempted to organize the deep sea fishers in Nova Scotia. The fish companies lobbied the government extensively, arguing that the workers on the vessels were "co-adventurers," not waged employees, and therefore not eligible for workers' compensation. One company, Zwickers, went to court over the issue, with the financial backing of the other major enterprises. In the case of *Re Zwickers; Re Application of Lunenberg Sea Products Ltd.* (1947), the Scotia Supreme Court found in favour of Zwickers, and the decision, in which fishers were defined as "co-adventurers," established a legal precedent. The government acted swiftly; in 1947, the revised Nova Scotia *Trade Union Act* specified that co-adventurers were not "employees." This change denied fishers the right to unionize. No attempt to organize deep sea fishers was made until 1968, when the United Fishermen and Allied Workers' Union tried to organize the Canso workers.

The 1960s saw important changes in the East Coast fishery. Gradually the federal government became interested in working conditions. In 1962, it set up the Canadian Coast Guard, including the Marine Casualty Investigations Division, and in 1964 it revised the *Canada Shipping Act*. For the first time, the government required engineers (1965) and masters and mates (1968) to have formal training and certification to operate deep sea vessels. It also surveyed the labour force to find out about working conditions in the deep sea fishing fleet (Proskie and Adams, 1971).

From 1965 to 1968, the Nova Scotia Commission Regarding Workmen's Compensation Inquiry looked into Part III of the *Workmen's Compensation Act*. All of the final reports, except the one presented by the Canada Labour Council, which represented the fishers' interests, recommended that fishers stay under Part III. During the inquiry, a series of sinkings of deep sea vessels alerted the public to the dangers in the industry (Winsor 1987, 75). These events, combined with the attempt by the British Columbia–based United Fishermen and Allied Workers' Union (UFAWU) to organize the Nova Scotia deep sea fishers, put additional pressure on the government to address this issue. Responding to this pressure, Commissioner Clarke looked at the experience of other provinces. In 1955, the British Columbia Workers' Compensation Board had become the first board to include fishers.[4] But the BC legislation limited the type of fishers covered and restricted the preventive measures and educational programs that the board could provide on behalf of fishers. Clarke concluded that the way the British Columbia Workers' Compensation Board had incorporated fishers into their program could be applied in Nova Scotia. He also noted that Nova Scotia remained the only province with a deep sea fishery that continued to exclude those fishers from workers' compensation benefits.

However, a few obstacles needed to be overcome before Clarke's subsequent recommendations could be put into effect: most importantly, the changing of the "co-adventurer" status of deep sea fishers. From 1968 to 1972 the UFAWU attempted to unionize deep sea fishers and fish plant workers. This represented a direct challenge to the Nova Scotia *Trade Union Act*, which denied fishers the right to join unions. The UFAWU's attempt failed and they left the province, but this labour struggle made the general public aware of the working conditions in the fishery.[5]

In 1971 the provincial government amended the Nova Scotia *Trade Union Act*, giving fishers the right to unionize. By 1972 deep sea fishers could join either the Canadian Food and Allied Workers (CFAW) or the Canadian Brotherhood of Railway and Transport Workers (CBRT). In 1972 the amended Nova Scotia *Workers' Compensation Act* included fishers in Part I of the Act.

During the 1970s the fishery declined due to shrinking fish stocks, the loss of frozen fish markets in the United States, lower company profits, and a series of strikes. The prevailing attitude in public policy of the day divided the fisheries into two categories: one that was economically viable (the deep sea fishery) and the other that supported the social fabric of fishing communities (the coastal fishery). In one sense this viewpoint accurately portrayed the situation: the income derived from coastal fishing hardly sufficed for survival, whereas the income earned in deep sea fishing ranged from modest to substantial, depending on boat ownership, job designation (crew or captain), and the current market value of fish (Davis and Thiessen 1986). Those economic advantages exacted a price, particularly in job satisfaction, family life, and community solidarity (Apostle, Kasdan, and Hanson 1985; Thiessen and Davis 1988; Binkley 1990b). Vessels sailed from fewer harbours and the crews came from further away. The discovery of offshore oil and the establishment of nearby industrial plants, such as Michelin Tire, reinforced companies' inability to recruit local workers in the 1970s. Both of these developments drew potential workers away from the deep sea fishery and created a shortage of experienced crew members. Companies had to recruit workers from distant villages and from Newfoundland.

The development of the 200-mile exclusive economic zone in 1977 increased the potential of the deep sea fishery. Not only did it allow for greater control and further exploitation of the resource by the

Canadian deep sea fleet, but it encouraged companies to expand their endeavours into the northern waters above the Straits of Labrador, into the Davis Strait and beyond. That expansion required improved technology. The necessary improvements, such as ice davits and reinforced hulls, increased costs and the companies overhauled their vessels gradually. However, the fleet could fish in these waters without those improvements by improvising partially satisfactory methods to deal with ice. For example, one such technique, "chaining off the warp"[6] could be used in lieu of ice davits, but it increased risks and resulted in a number of accidents.

In 1983 the federal government formally recognized the economic crisis in the fishery with the appointment of Senator Michael Kirby to head a task force on the Atlantic fisheries. In the report *Navigating Troubled Waters: A New Policy for the Atlantic Fisheries,* Kirby called for the reorganization of the Atlantic fishery, including the restructuring and refinancing of the major fish companies, industrialization of the deep sea fleet, the elimination of part-time fishers and the introduction of factory freezer trawlers. The subsequent restructuring produced an economic improvement in the deep sea fishery, at the price of a further decline in the coastal fishery. The government, committed to this policy, accepted most of Kirby's recommendations; in the mid-1980s it granted three licences for deep sea factory freezer trawlers to companies based in Atlantic Canada.

The 1970s and 1980s saw a steady improvement in the level of safety on vessels. These changes came about partly through pressure from the unions – particularly the CFAW and the Newfoundland-based unions – in the form of strikes. But the increasing use of scientific management procedures, along with the restructuring and refinancing of the major fish companies, also contributed to this change. Nevertheless, during the late 1970s and early 1980s a series of sinkings and serious accidents continued to plague the industry, and workers' compensation premiums rose dramatically. Once again, as in 1928, the companies lobbied the provincial government for relief. As a result, the board assigned premiums based on individual companies' performances and not on industry-wide performance. Companies now had a substantial incentive to improve safety and decrease costs. The more progressive companies recognized the cost-effectiveness of better maintenance, increased safety, and fewer accidents. Those companies, specifically National Sea, Scotia Trawlers, and McLeod's (now Clearwater Fine Foods), introduced new safety procedures. They emphasized training for officers and crews, developed routine maintenance programs, and hired managerial staff to oversee safety concerns, all in the hope of reducing down time due to accidents and mechanical breakdown.

Currently within the industry, training programs at all levels have been enhanced. The requirements for Master Mariner, Mates and Bosun Certificates (tickets) have been upgraded and a Master Mariner First Class (Fishing) has been introduced. Marine Emergency Duties courses (MEDI, II, and III) have been developed and offered to crew members as well as officers. These programs include training in first aid, fire-fighting, survival at sea, and lifesaving appliances. The provincial and federal governments, as well as individual companies, offer additional courses in first aid. Fishers are encouraged to upgrade their tickets, as well as their first aid, fire-fighting and MED qualifications, routinely. Some companies pay for the courses, others offer them on company time. One company currently requires – and pays for – all their captains to take the Master Mariner First Class course. But some companies still do not support these activities.

The Atlantic Record Book Plan, sponsored and developed by the Large Vessel Association, offers new recruits a deckhands' course, which covers basic working practices at sea. Although the program was slow to get off the ground because some of the more cynical union leaders saw it as a way to exclude "troublemakers," most men now entering the industry take this course.

The federal government now requires additional safety equipment on fishing vessels. Beginning in 1985 all vessels over 150 tons (i.e., most vessels over 100 feet) had to carry survival suits. Most companies require fishers to wear hard hats, flotation vests and steel toed shoes when working on deck and hauling back and shooting away the fishing gear. Requirements for fire-fighting equipment have become more stringent, and some companies require monthly fire and boat drills.

During the 1980s the dispute over jurisdiction and responsibility for the health and safety aspects of the fishery escalated. In 1976 the Worker's Compensation Board of British Columbia introduced a number of regulations intended to put British Columbia fishers under the *Canada Labour Code* regulations for land-based workers. The federal Department of Transport, which oversees the Canadian Coast Guard, took the British Columbia Workers' Compensation Board to court. The Supreme Court of Canada ruled in favour of the Department of Transport, stating that the *Canada Shipping Act* oversaw the fishery. The Department of Transport did not introduce any corresponding legislation for marine workers.

In 1984 the Nova Scotia Government appointed the Committee on Occupational Health and Safety, chaired by Dr. Tom MacKeough, former Nova Scotia Minister of Labour, to assess the legislation pertinent to provincial occupational health and safety. Once again a commission ignored the special concerns of the fishery. This time two unions, the Maritime Fishermen's Union (primarily a coastal fishers' union) and the CFAW, made presentations. A number of submissions were made by industry as well. Yet the report (Nova Scotia, Committee 1984) did not comment on the fishery, assuming, it would seem, that it was out of the committee's jurisdiction, since fisheries came under the federal government.

In 1987, when the prime minister and the ten provincial premiers met in Winnipeg, they discussed occupational health and safety in the fishery. Their concerns centered on the high number of accidents and injuries, the lack of basic "rights" of workers on fishing vessels (i.e., the right to refuse unsafe work) compared to land-based workers, and the exclusion of small vessels from any protection or support. They set up the tri-party Committee on Occupational Safety and Health in the Fishing Industry with representatives from government, industry, and labour. The committee's *Report* (1988) emphasized training and education as ways of improving the work environment and promoting safer work practices. The question of which government agency should control the regulations and their enforcement remained unresolved. The committee did recommend that the Departments of Transport and Labour work more closely together.

Two additional studies appeared at about this time, one by R.J. Gray (1987) and the other by the Canadian Coast Guard (Canada, Coast Guard Working Group 1987). In *An Examination of the Occupational Safety and Health Situation in the Fishing Industry in B.C.,* Gray argued that the Department of Labour, not the Department of Transport, should regulate working conditions on fishing vessels. *A Coast Guard Study into Fishing Vessel Safety* argued for the maintenance of the status quo. This struggle over jurisdiction, coupled with the *Ocean Ranger* disaster in 1982, forced the Department of Transport to pay more attention to the needs of fishers and to revise their regulations along the lines proposed by the British Columbia government in the late 1970s. However many people in the industry, especially union leaders, did not think these measures went far enough.

On 2 April 1987 the Department of Labour (Labour Canada) took over responsibility for the "occupational safety and health of employees employed on ships registered in Canada or on uncommissioned ships of Her Majesty in the right [*sic*] of Canada and employees employed in the loading and unloading of ships" (Canada, *Canada Labour Code, Marine Occupational Safety and Health Regulations,* SOR/87-183). For the first time a government agency had direct responsibility for marine industrial safety and health. Little has changed since then.

CHAPTER THREE

Working Conditions

Fishing enterprises do not operate in a vacuum. They are integrated into the broader socio-economic environment that affects the working conditions of fishers both directly and indirectly. Of the many factors involved, this discussion focuses on a few that have been historically significant for the Nova Scotia fishery: resource management policies (enterprise allocations), market pressures (e.g., to increase quality), labour legislation (the Nova Scotia *Trade Union Act*), and socio-economic characteristics of fishers' home communities (e.g., work opportunities for wives).

Working conditions consist of the circumstances under which workers labour; job satisfaction consists of the attitudes and feelings of workers toward and about their work. These related and complementary concepts serve as a basis for understanding variations in work organization within the fishery. This chapter describes working conditions in the Nova Scotia deep sea fishery and examines factors affecting those conditions at two levels: the impact of local factors on the immediate work environment, and the impact of external socio-economic environmental factors on each enterprise. After describing the work organization and working conditions of the fishery at both levels of analysis, I illustrate those conditions with an account of a typical fishing trip. In the next chapter I analyse our survey data to break out information on job satisfaction.

The physical environment, the level of technology, and the social organization of the particular enterprise influence working conditions on fishing vessels. Fishing, unlike almost any other industry, involves a hunt for a mobile resource – and one that must be captured in a hazardous environment that cannot be controlled. The North Atlantic is one of the cruellest seas, especially during the winter months when ice and fierce gales disrupt fishing and shipping. But even on the

Immediate environment	Sectors of the enterprise		
Sea environment	Harvesting (marine resource)		
Level of technology		MANAGEMENT	EXTERNAL ENVIRONMENT
Work organization	Processing (product)		
	Marketing		

Figure 3.1
Model of factors affecting the organization of deep sea fishing enterprises

calmest day the industrial site of production is constantly moving. As one captain pointed out: "Fishing is not an easy way to work. You work on land in a factory, the factory floor doesn't move. You're on a boat and your floor is moving, the stuff on it is moving and it's just a matter of staying out of the way sometimes" (Well Workers' Interviews, Captain). In response to this challenge, the industry relies on technology to mitigate obstacles, to protect fishers from the perils of the sea, and to locate, catch, process, and store the marine resource.

The type of technology used depends on the type of resource available (e.g., scallops, groundfish) and its location, as well as the level of capitalization of the enterprise and the safety measures deemed necessary. The organization of work on the vessel depends directly on the technology used and the degree of vertical integration of the enterprise, and indirectly on all the factors affecting the choice of technology (see figure 3.1).

The rapid industrialization of the deep sea fishing industry in Nova Scotia has resulted in wide variations in work organization.[1] Other industries have undergone similar expansion (cf. Child 1975, 1984; Child and Partridge 1982; Hackman, Pearce, and Wolff 1978; Kelly 1982; Reimann and Inzerilli 1979; Rowbottom and Billis 1977; Salaman 1979), and many parallels could be drawn between those experiences and the ones in the deep sea fishery. Comparative ethnographic material on work organization of the industrialized North Atlantic deep sea fishery[2] has been collected by other researchers, notably Andersen (1972, 1979), Andersen and Stiles (1973), Aubert and Arner (1958), Horbulewicz (1972), Neis (1987), and Tunstall (1962).

A typical deep sea fishing enterprise incorporates the harvesting, processing, and marketing sectors into a single enterprise. The harvesting sector catches and transports fish of an adequate quality to the processing plants within a reasonable time. It may also clean and gut the fish. The processing sector provides primary, and in some cases secondary, processing of the fish. The marketing sector sells the product, determines products' present and future market potential, and advises management about those issues. Management oversees the enterprise, determines what kind, where, and how much fish will be caught and where they will be transported for processing. Management also ensures the integration of data from each sector of the enterprise into its overall operation.

THE IMMEDIATE WORKING ENVIRONMENT

The fishing grounds off the shores of Atlantic Canada have been described as some of the world's most bountiful. John Cabot (in Innis 1954, 11) noted that "the sea there is swarming with fish, which can be taken not only with a net but in baskets let down with a stone, so that it sinks in the water." In *Captains Courageous*, Kipling (1982, 91) described the resources of the Grand Banks of Newfoundland: "The sea round them clouded and darkened, and then fizzed up in showers of tiny silver fish [caplin], and over a space of five or six acres the cod began to leap like trout in May; while behind the cod three or four broad gray-black backs [whales] broke the water into boils."

The fishing grounds off the shores of Atlantic Canada extend from the Gulf of Maine in the south to the Davis Strait in the north, west from the Gulf of St. Lawrence to the eastern waters off the Newfoundland coast. The seaward boundary includes the shoals and banks running from Georges and Browns Banks off the south-eastern shores of Nova Scotia to the Grand Banks off Newfoundland and follows the continental shelf to the hundred-fathom line, which in some places stretches more than three hundred nautical miles offshore (see figure 3.2). The combination of the cold Labrador current and the broad continental shelf make this area a fertile feeding ground for many species of fish and other marine plants and animals. Of the numerous marine resources available fishers most commonly harvest lobster, groundfish (bottom feeders, such as cod, pollock, and flounder), herring, salmon, shrimp, molluscs, and previously underutilized species such as dogfish and squid (see table 3.1).

With two exceptions, fishers from the provinces of New Brunswick, Nova Scotia, Prince Edward Island, Quebec, and Newfoundland have exclusive rights to exploit the marine resources within the two-hundred-mile Canadian economic zone. France claims the right of fishers from the French islands of Saint-Pierre and Miquelon to fish within the two-hundred-mile economic zone around their islands, and the United States claims the right of its fishers to fish off Georges Bank

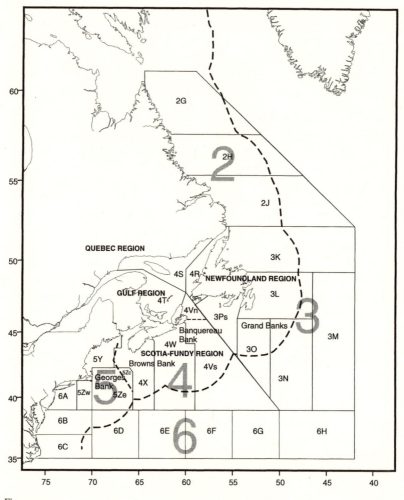

Figure 3.2
Map of Atlantic Canada showing fishing zones and the 200-mile limit
Source: Canada, Department of Fisheries and Oceans, Communication Centre, Halifax, 1994.

in the Gulf of Maine. Both these claims have gone to international arbitration and the parties have achieved mutually satisfactory settlements. Joint economic ventures between Canadian companies and those from other countries (e.g., a Scandinavian-Canadian shrimper working off Labrador) also fish in these waters.

Effective management of marine resources has been a long-term goal of the Canadian state. The government separates the fishery into sectors based on species, location, and level of technology. Each sector

Table 3.1
Summary of Nova Scotia fishery products by main species, 1986

Species	Quantity in tonnes	Value in $1,000s
GROUNDFISH	174,579	566,983
Cod	68,857	258,952
Haddock	20,801	72,219
Pollock	19,527	51,975
Hake and cusk	6,176	14,247
Ocean perch	11,399	38,833
Catfish	857	1,498
Halibut	2,403	20,168
Turbot	101	296
Flounder and sole	8,112	24,405
Others	36,346	82,390
PELAGIC AND OTHER FINFISH	36,728	81,380
Herring	30,267	65,924
Mackerel	4,170	3,372
Alewife	1,205	686
Eel	2	4
Salmon	53	277
Smelt	12	33
Others	1,019	11,084
MOLLUSCS AND CRUSTACEANS	21,304	220,565
Clam	197	693
Oyster	51	151
Scallop	4,855	74,108
Squid	–	–
Lobster	13,112	128,646
Crab	792	3,637
Shrimp	2,152	13,230
Others	145	120
MISCELLANEOUS	12,071	881
Total	244,682	869,809

Source: Canada, Department of Fisheries and Oceans (1991, 82).

has a seasonal quota or allocation assigned to it. Once the quota is met, the fishery is closed. The fishery has two major divisions – the coastal or "inshore" fishery, and the deep sea or "offshore/midshore" fishery.[3] Both fisheries have limited entry. Small boats, usually under forty-five feet and equipped with simple technology (i.e., traps, nets, and lines), characterize the coastal fishery. They exploit a wide range of marine resources, and usually make daily fishing trips of less than ten miles offshore. The deep sea fishery uses larger vessels, forty-five feet and over, that employ seines, trawls, and drags to fish primarily for groundfish and scallops. Each fishing trip lasts for a week or more and ships travel many miles from their home port. Vessels between forty-five and sixty-five feet fall between the two sectors. Since the size of crew and trip length also lie midway between the other two sectors, I refer to this category as the "midshore" fishery. This study focuses only on the deep sea fishery.

All of these vessels work under a quota system based on season, species, and gear type. In the midshore fishery, all vessels compete for the quota in their sector, although this system will gradually be replaced by an individual transferable quota (ITQ) system. For the deep sea scallop and trawler fleets, the government, in consultation with all of the enterprises in these sectors, assigns an individual quota – an enterprise allocation (EA) – to each company or enterprise. These quota systems maintain the sense of challenge and independence for workers in the midshore fleet but diminish it for workers in the trawler and the scallop fleets.

THE TECHNOLOGY

The type of vessel and other forms of technology that fishing companies and fishers use depend on the type of marine life they want to harvest. The two major types of technology employed in the Nova Scotia deep sea fishery are trawls and drags. The trawler fleet catches groundfish using a ground or a midwater trawl. The scallop fleet dredges the ocean bottom for scallops.

Trawlers

Stern trawlers, usually large vessels ranging in size from 65 to over 250 feet, can be recognised even at a long distance by a metal arch or "gallows" that spans the deck at the stern. Stern trawlers "shoot and haul" their gear from the stern of the vessel; side trawlers, now obsolete, shot and hauled their gear from the side. The *Cape North*, the first Canadian factory freezer trawler, guts, cleans, and freezes fish on board. On all other trawlers the fishers only gut, wash, and store the fish on ice, either in pens or in rows of plastic containers in the refrigerated hold. Figure 3.3 illustrates the layout of a typical Nova Scotia trawler. These vessels carry highly sophisticated fish-finding and navigation equipment, such as lorans (*lo*ng-*ra*nge *n*avigation *s*ystem) and radar, radios, automatic pilots, FAX machines, and modern safety and fire-fighting equipment.

The trawler gets its name from the large baglike nets that the vessel tows (see figure 3.4). The two large side wings of the net narrow to a

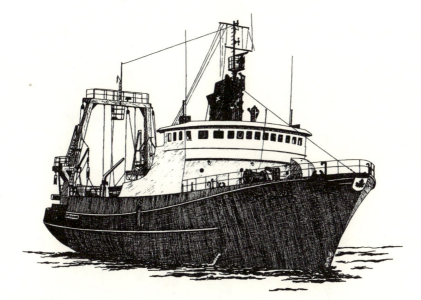

Figure 3.3
Modern stern trawler
Source: Nova Scotia, Department of Fisheries (1979, 25).

cone-shaped area, the *belly*, which ends with a large knot, the *cod end*. As illustrated in figure 3.4, the net is towed by two *warps* (wire cables) fastened to the *doors* (*otter boards*), which are attached both at the top and bottom of the net on each side by *bridles* or ground cables. The mouth of the net is held open horizontally by the doors, and vertically by a series of floats stitched into the top of the net and by weights attached to the bottom of the net. As the net passes through the water, it collects all material and marine life that is larger than its mesh.

Two types of trawls are used – the ground trawl and the midwater trawl – depending on the type of fish sought and the location of the fish in relation to the bottom of the ocean. The ground trawl, the commoner type, has a series of steel balls, or rollers, attached to the bottom of the net so it can "roll" along the sea bottom and harvest fish close to the ocean floor. Trawlers use the midwater method to harvest fish that congregate above the sea bottom. Because the trawl is not pulled along the ocean floor, it does not have rollers. A depth finder, called a transducer, is sewn into the top of the net so that the depth of the net can be controlled.

Trawling requires a high degree of teamwork (see figure 3.5). The work can be divided into three separate parts: the *shooting* of the trawl, the *hauling back* of the trawl, and the processing of the fish. Shooting the gear consists of throwing the wings, belly, and cod end of the net

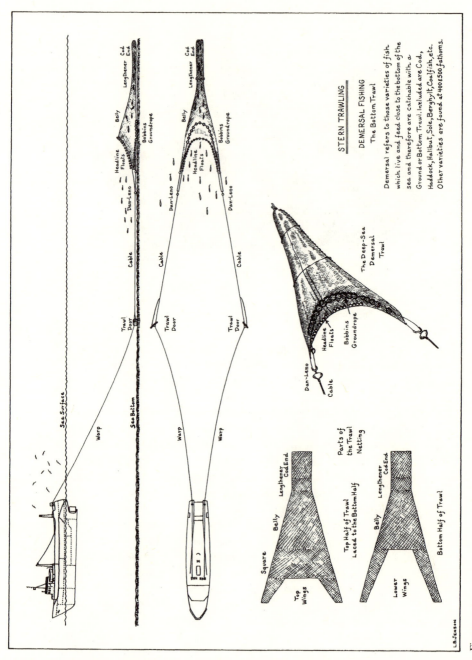

STERN TRAWLING

DEMERSAL FISHING
The Bottom Trawl

Demersal refers to those varieties of fish
which live and feed close to the bottom of the
sea and therefore are catchable with a
Ground or Bottom Trawl. Included are Cod,
Haddock, Halibut, Sole, Berghylt, Coalfish, etc.
Other varieties are found at 400 to 500 fathoms.

The Deep-Sea
Demersal
Trawl

Sea Surface

Warp
Sea Bottom

Cable
Trawl Door
Dan-Leno
Headline Floats
Belly
Lengthener
Cod End
Bobbins
Groundrope

Cable
Trawl Door
Warp
Dan-Leno
Headline Floats
Belly
Lengthener
Cod End
Bobbins
Groundrope

Dan-Leno
Cable
Headline Floats
Bobbins
Groundrope

Square
Belly
Lengthener
Cod End
Top Wings

Parts of
the Trawl
Netting

Top Half of Trawl
Laced to the Bottom Half

Belly
Lengthener
Cod End
Lower Wings
Bottom Half of Trawl

Figure 3·4
Stern trawling

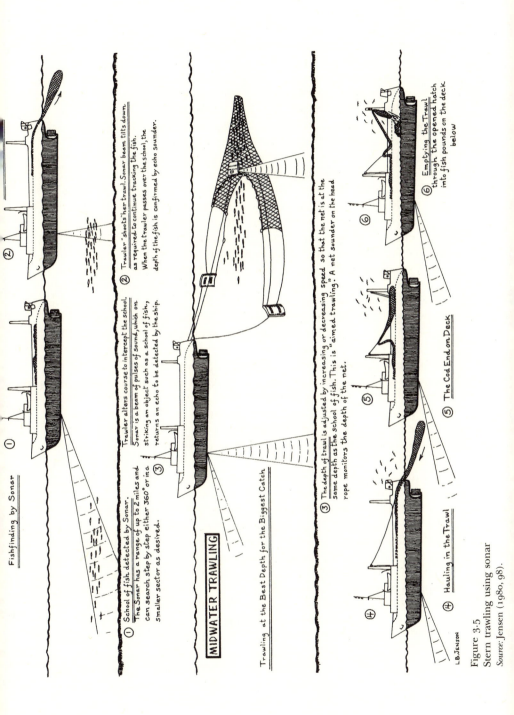

Fishfinding by Sonar

① School of fish detected by Sonar. Trawler alters course to intercept the school. The Sonar has a range of up to 2 miles and can search step by step either 360° or in a smaller sector as desired.

Sonar is a beam of pulses of sound, which on striking an object such as a school of fish, returns an echo to be detected by the ship. ③

② Trawler "shoots" her trawl. Sonar beam tilts down as required to continue tracking the fish. When the trawler passes over the school, the depth of the fish is confirmed by echo sounder.

MIDWATER TRAWLING

Trawling at the Best Depth for the Biggest Catch

③ The depth of trawl is adjusted by increasing or decreasing speed so that the net is at the same depth as the school of fish. This is "aimed trawling." A net sounder on the head rope monitors the depth of the net.

④ Hauling in the Trawl

⑤ The Cod End on Deck

⑥ Emptying the Trawl through the opened hatch into fish pounds on the deck below

L.B. JENSEN

Figure 3.5
Stern trawling using sonar
Source: Jensen (1980, 98).

off the stern while the vessel turns in a wide arc, playing out the ground cables. The doors are attached to the ground cables and then to the warps. The warps are then released, and as they are played out markers indicate how much cable has been extended. When the desired amount of warp has played out, the brakes on the winches are engaged and the net is set. The net is usually set for one and a half to two hours, depending on the abundance of fish, and then the net and its contents are hauled in. It is important to haul the catch in frequently. Too large a catch can lead to torn gear, resulting in lost catch and time wasted on repairs.

Hauling back the trawl begins with the vessel swinging in a slight arc, while maintaining its speed. Once the vessel has come about, the warps are hauled in, and wound on the winch drums. The depth markers on the warps indicate the amount of cable that has been hauled in. The winch is slowed once the hundred-foot marker has been reached. The first thing to appear out of the water are the doors. These are hauled in, unshackled, and chained to the gallows, the large steel arch at the stern of the vessel. At this point, the cod end and the belly of the net with its catch may be seen floating on the surface off the stern. The ground cables are now hauled in by the winch. After the wings of the net have come out of the water and have filled the "horseshoe" area of the deck around the stern, the winch is stopped. The ends of the quarter ropes are then untied from each wing and are carried along the leads to the winch, taken around the niggerhead [sic] (auxiliary winch) and hauled in. When the bag has been pulled in, the winch is stopped. The bag is now suspended from the gallows over the ramp that leads to the fish rooms or processing area below deck. The ramp is opened, a member of the crew unties the cod end, and the fish slide down the ramp into the storage area. The cod end is retied and the whole process is repeated.

Once the net has been shot again, the crew processes the fish. In the fish rooms the fish are sorted and processed either by machine or by hand, depending on the species. The gutted fish are washed and sent by conveyor belts to the refrigerator hold. There are two basic methods of storing fish, pens and boxing. In the pen system, the hold is divided into square pens (usually four to six feet square), and fish of a single species are mixed with ice and placed in the pen. The pens are built up vertically with aluminum slats, and aluminum slats are also placed horizontally across the pen every three to four feet, before the next load of fish are delivered. This process takes damaging pressure off the fish. The boxing system uses plastic boxes that hold either 70 or 110 pounds of fish each. Individual species of fish are mixed with ice and placed in the boxes, which are then stacked one

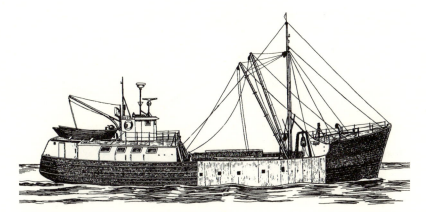

Figure 3.6
Offshore scallop dragger
Source: Nova Scotia, Department of Fisheries (1979, 20).

on top of the other in rows in the hold. This method is replacing the pen system even though it is more labour intensive, because it ensures a higher quality of fish.

Scallopers

Scallopers are the other major type of deep sea vessel in Nova Scotia. These boats are usually between 65 and 150 feet, and they fish scallop beds usually only a few hours' sail from home port – particularly Georges, Browns, and Banquero Banks, and the banks of the Bay of Fundy (see figure 3.2).

Many of the vessels used in scalloping are converted wooden side trawlers, although there are a few new boats designed and built specifically for scalloping. Scallopers can be distinguished from trawlers by the heavy sheathing on the sides that protects the wood from the metal drags, and by the long booms located on each side of the vessel, which are used to hoist the drags when shooting or hauling away (see figure 3.6). There are no aft (stern) gallows as on trawlers; instead, there are two side gallows. The deck aft is enclosed, forming the shucking house, where the crew shuck the scallops while protected from the weather. The vessel also carries navigational aids, safety, and fish-finding equipment, as well as other basic telecommunications equipment.

The basic routine of fishing for scallops is similar to fish trawling. The fishing gear – *drags*, or rakes – consists of a heavy metal frame and a bag made of steel rings (see figure 3.7). Scallopers tow two drags (rakes), one on each side of the boat, and the deck arrangement is

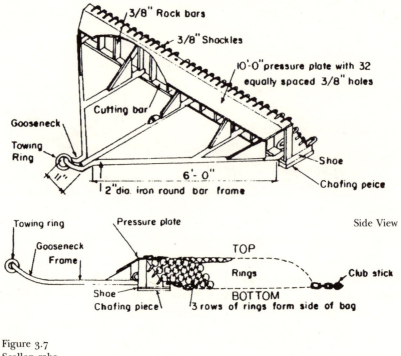

Figure 3.7
Scallop rake
Source: Barney and Carberry (1987, 40).

such that both drags can be handled at the same time. The deck has
a raised hinged area on each side called a dump table, where the
scallops are picked out of the other debris. Once the scallops have
been removed, these tables are raised and their contents dumped over
the side. As with trawling there are three separate stages: shooting
away, hauling, and processing.

At the beginning of shooting away, the frame of the rakes rests on
the rails and each bag hangs overboard, held in place by the *knock out
block* and by the *stay chains*, which are restraining chains bolted to the
deck and hooked into the sides of the rake's bag. When the crew is
ready to shoot the rakes, the boat speeds up, the warps are attached,
and the chains removed. The skipper then sounds the horn, and when
the vessel falls down in the swell, the crew knocks out the block and
the winchman allows the warp to play out. The rake falls overboard
when the block is removed. This procedure is carried out on each side
simultaneously. The winchmen play out the required amount of warp,
the brakes are set on the winch, and the vessel tows the drags for
twenty to thirty minutes at a slower speed.

L.B.JENSON

Figure 3.8
Boarding a scallop drag
Source: Jensen (1980, 90).

When the tow is over, the engine is slowed down and the rakes are hauled in simultaneously. Once the twenty-five-fathom warp marker is reached, the crew comes out of the shucking house and makes ready to board the rakes, with one man at each winch and gallows. When the rake surfaces, the winch is disengaged and the brake applied. A crew member attaches the boom tackle hooks into the towing rings at the front of the rails. The rakes come up with their backs against the vessel, and they must be reversed in order to bring them on board.

Figure 3.9
Shucking scallops
Source: Bourne (1964, 47).

To do this, a stay chain is hooked into the after (lower) end of the rake's frame, the winch brake is released, and the weight of the rakes is taken up by the boom's tackle. The drags now are easily turned so that their bellies are against the vessel and they can be brought aboard once the stay chains are cast off.

As illustrated in figure 3.8, as each rake comes on board the restraining chains are re-attached. The drags are lifted high enough to clear the rails and as the vessel rolls the drags swing on board onto the dump tables. The boom tackle is hooked into the rings attached to the *clubstick* (metal bar at the end of the rake's belly) chains. The cargo hoists are re-engaged, the clubsticks raise the bags, and the contents of the bags spill onto the dump tables. The clubsticks and bags are lowered onto the side so the frames rest on the rails ready to shoot away again as soon as the skipper is ready. Once the rakes have been shot away, the crew sorts through the catch, according to size and species. Scallops with shells larger than 3¾ or 4 inches are collected, and once those scallops have been removed the trash is dumped overboard.

Processing the scallops is done at sea and by hand. Figure 3.9 shows how scallops are shucked. Assuming the shucker is right-handed, he holds the scallop in his left hand with the hinge toward his palm and the flat shell upwards. Holding his knife in his right hand, the shucker inserts it forward and upward along the inner face of the flat shell, entering just behind the corner of the hinge closest to his body, nearest to his baby finger. He forces the blade downward and backwards in a semicircular motion toward himself, severing the attachments of the "muscle" or meat and the "rim" or viscera from the flat shell. He hooks the point of the knife downward under the thick muscular mantle and away from himself, pressing his thumb against the shell and clamping the mantle edge between it and the knife. Now the shucker lifts the knife upward and toward himself, lifting the upper valve of the shell and the whole rim away, leaving only the meat attached to the lower valve in his palm. He then scrapes out the meat

Table 3.2
Summary of working conditions in the Nova Scotia fisheries

Working conditions	Vessel and gear types		
	Midshore	Scalloper	Trawler
Gear	Mixed gear	Scallop drags	Ground trawl Midwater trawl
Vessel size	45–64 feet	65 feet and over	65 feet and over
Level of capitalization	Craft	Mixed	Industrial

into the shucking pail. An experienced shucker can complete this procedure in a matter of seconds. Once the scallops have been shucked they are washed in salt water and put in cotton bags that hold about thirty-five pounds. They are covered completely in ice, and stored in either plastic containers or pens in the refrigerated hold.

Defining the Fleet

For the survey I divided the deep sea fishing fleet into three groups based on the different technology used, specifically the vessel and gear type. These groups were

1 vessels sixty-five feet and over with groundfish trawls (trawler fleet);
2 vessels sixty-five feet and over with scallop drags (scalloper fleet); and
3 vessels over forty-five but under sixty-five feet of mixed gear type (midshore fleet).

Throughout the book, these fleets are called scalloper, trawler and midshore respectively (see table 3.2).

The trawler fleet on its fishing grounds may be as far as a two-day steam from home port. The companies will send a vessel to a specific place, with instructions about what type of fish and how much of each species to catch. This offshore fishery operates year round, and workers are usually away for ten to fourteen days with two days off between voyages. Crews range from ten to sixteen workers, who work on highly mechanized vessels where both the catching and processing are done mechanically. The worker has little control over the pace of the work. Although the crew is divided into watches (shifts), these shifts are frequently broken (extended) and men work gruelling hours. By reputation, this is the hardest fishery. This is how one trawlerman describes his work:

My job – I'm down in the hold icing fish. In rough weather when the boat's rolling, you got a hard job down there just to stand up let alone try and stand up and do your work at the same time. We got them aluminum pen boards. They're slippery … When you're down in the hold, you gotta shovel all the fish up into the pen. If the boat rolls, all that comes down on you. You shovel up the ice and the fish slides down on you. Eighteen hours in the hold shovelling ice is bad enough without that. It don't do you no good. You get sore legs, 'specially down around there [calves and ankles] from standing up for all that long shovelling. I don't bother to use liniment. I just let it go on its own. You get over it after a while.

You're suppose to get six hours sleep. By the time you get to sleep it seems like it goes so fast. But sometimes you don't get six hours. Like the other trip there, we was up for twelve hours. They turned in for three hours. Then we was up again for another eighteen hours. We had lots of fish. It's in the contract, you're only suppose to work eighteen and six, but everybody's got different ways of working, I guess. After eighteen hours work you're not much good anyway [chuckles]. I tell you, you gets tired after doing that for awhile. You'd get down in the hold quick for a higher price, but you don't get it [chuckles]. (Well Workers Interviews, Deckhand)

The scallop fleet usually dredges the scallop beds only a few hours away from home port. In order to spread out the quota, companies try to space out vessels' sailing times. Workers on scallopers are away for five to eight days and they have three to four days off after each trip. They do not fish in late December or January. Depending on the size of the vessel, crews consist of ten to fourteen men who work on a watch system. The living conditions on scallopers are cramped and notoriously poor. Although the dredges are mechanized, the workers can control the speed of this work. All of the crew are involved in dredging and processing the scallops, including shucking. Here is how one scalloper described his job:

I've been working on scallop draggers for twenty-six years – going on twenty-seven. Always been a deckhand. Never done anything else. This boat I'm on now is different. But the boats I was on before, when you went out, you stood eight on and four off for twelve to fifteen days. There was no walking with a basket of scallops or anything like that, you run with a basket of scallops. That's just the way you was goin' for fourteen days. You just wore yourself out. If the skipper seen that you wasn't hurrying at everything you done, when you got to the wharf you took your bag and that was it – you got put ashore. But now it's different, it's a little easier. Like I say, us scallop fishermen, we put up with that for years. But it is getting a little better.

It takes about five minutes to haul back and shoot away. In six hours, we'd haul back around ten times, if nothing goes wrong. You'd stand two six-hour watches and you'd haul back about eighteen or twenty times. You have one fella running the winch, one fella working the after end, and one fella working the forward end. If you got a spare man, he's back shucking. Then, the next time that you haul back the spare man takes his turn out on deck and the other fella stays back shucking. But the people that shucks are the people that works. You're running back and forth handling gear and picking up scallops, all the time. Unless you happen to get a slack tow. Then you [sit] in the galley for a little while and pick at the cook [laughs]. You sleep where you want to or where you can, and you work where you're put. (Well Workers Interviews, Deckhand)

The midshore fleet uses both scallop drags and trawls. These ships usually fish only a few hours from their home port, and trips tend to last five to seven days. They return to their home port as soon as they have a full catch, which they sell to local processors. The crew are paid on the share system, where the net proceeds are divided among the eight or ten crew members.[4] These vessels are not as highly mechanized as the trawler and scallop fleets, and the fishers have more control over the pace of the work. Since the work is not divided up into regular watches, the work rhythm reflects the catch size. According to deep sea workers, this fleet has the best working conditions.

THE ORGANIZATION OF FISHING ENTERPRISES

Different organizational settings influence how workers understand and describe their work (Silverman 1970; Silverman and Jones 1976). Norr and Norr (1978, 163, 169; 1974) argued that variations in work organization within the modern fishery depend primarily on increased levels of capitalization, which lead to control of fishing gear by nonfishers. They suggested that where fishers retained control of the means of production, work organization would have rational administration and nonbureaucratic structures. Similarly, where nonfishers control the means of production, there would be more bureaucratic structure, with a separation of major decision making from the direct influence of fishing conditions and crew interests.[5]

Traditionally, the organization of fishing enterprises has been divided into two types based on the levels of capitalization.[6] I define these groups as follows:

1 *Craft:* an independent owner/operator or a small company based in a single community that employs nonunion people from that community.

2 *Industrial:* a medium to large company based in a single community that usually employs people who come from many different com-

munities and who are usually union members. These enterprises may be vertically integrated companies and may have locations in more than one community.

Craft and industrial enterprises have structural similarities. Both types are capital-intensive and use similar gear and technology to catch fish. They draw on the same workforce to run their vessels. They differ importantly in work organization, with the crucial difference lying in control over the means of production. This, in turn, determines the management methods of the enterprises and the levels of bureaucracy necessary to run them (cf. Clegg and Dunkley 1980; Pfeffer 1981).

Work organization profoundly influences the structure of the fishing enterprise, and Norr and Norr characterize the organization typical of a fisher-controlled (craft) enterprise (1978, 169). The Nova Scotia coastal fishery shares these characteristics. The craft enterprise is usually owned and operated by a single individual or a small company. Although it is not part of a vertically integrated company, it may sell its catch to such a company, or directly to a broker. The organization of the craft enterprise is not managed through an extended personnel and production system, nor is this a family-based fishery.[7] The operator/owner controls the enterprise directly, without any outside intermediaries. Crews are recruited from within the community, and dissolve by informal agreement.[8] Each member has a personal set of obligations to the captain and other crew members. There is a spirit of egalitarianism – they feel they share a common task. After the owner/operator has deducted the trip's expenses, including such costs as mortgage and loan payments on the boat, each person receives a predetermined share of the net proceeds. The underlying premise in this system is that all take risks in the voyage, and all reap the profits of the venture. This is the rationale of the co-adventurer system.

The work schedule of these vessels is sensitive to a variety of influences – individual, family, economic, and weather – reflecting resource and community rhythms and schedules. Decisions on the vessels are based on discussions among the crew members, with the skipper making the final decision. The crew works in common to process the fish and do other tasks. The quantity and quality of the fish caught determine the work schedule on board. With this kind of flexibility, workers find it easier to plan participation in family events and community activities. This type of organization strengthens the social bonds among its workers by giving them a strong sense of community identity, and it promotes integration of crew members from the same community. They share common experiences, goals and values, and workers have a sense of personal control over their working conditions.[9]

Although deep sea fishing enterprises are industrializing rapidly, they have traditionally been craft enterprises. Today no deep sea enterprise is truly a craft enterprise; however, the midshore fishery retains some of those characteristics. The industrial enterprises emphasize formal training, credentials, and hierarchy. There are formal distinctions of authority, and there is little consultation across status levels. The crew is not involved in decision making. Although these enterprises also encourage the crew's teamwork and specialization in particular jobs, management officials at the company's home port recruit and screen workers. They hire workers for particular positions because they have specific skills, rather than for their general suitability. They choose crews from a list of available unionized workers according to their qualifications, training, and sea rotation, not personal preference. Although captains of vessels have the final say in who crews for them, this prerogative represents more of a veto power than a free hand in picking and choosing shipmates.[10]

*Work Organization in the Nova Scotia
Deep Sea Fleet*

In Nova Scotia there are three companies based in the Atlantic region and seven based in Nova Scotia that share common characteristics and management procedures. All vessels have corporate ownership. Even the nominally "owner-operator" vessels have corporate money invested in them.

These vertically integrated, year-round operations own deep sea fleets that supply primary, and in some cases secondary, processing plants on shore. Driven by market demands for specific types and qualities of fish products, companies develop and manage plants to meet these needs. The demands of the market bear on the plants, and the plants' requirements are transmitted to the vessels where the fish are harvested. Management control of plants is similar to that in other land-based enterprises. But controlling vessels at sea is difficult, and controlling the harvesting of fish is more difficult still. Like most industrialized land-based enterprises, these companies utilize management techniques that include modern accounting methods, cost efficiency programs, and time budgeting. Although primary production takes place at sea, chasing a mobile resource in an uncontrolled environment, companies attempt to transfer land-based management strategies to sea-based production.

The organization of workers into a social pyramid on these vessels reflects an industrial model. The captain (usually called the *Master*) is at the top. Officers (mate, engineers, bosun, and cook) are in the middle. The crew (trawlermen, deckhands, and learners) are at the bottom. There is a basic social division, as well as a division of labour, between the officers and crew. The work is divided into jobs with defined activities, and each job has a classification structure, position description, and graded pay distribution (number of shares).

Workers are unionized, and their work schedule depends on the needs of the company.[11] Watches on board are usually six hours "on" and six hours "off." This schedule continues uninterrupted through-out the trip except when the fishing is heavy. Then the crew work through their off shift, sometimes for as long as eighteen hours.

Captains and mates are managers at sea, holding supervisory posi-tions on board the vessel. They also exercise their traditional roles as persons responsible for finding the fish and navigating the vessel. In order to assert central control, companies maintain communications through radio contact between the vessels and the company's central office. Each day, at a specific time, the "daily hail" takes place. The company radios its managers at sea requesting information on their position, catch size, and any problems they have encountered. Man-agement may also pass on additional instructions, or advise the captain of a change in plans.[12] Vessels may be rerouted to locations where a particular type of fish can be caught, or vessels with full loads may be sent to plants where the fish are needed, and not necessarily to the vessel's home port. This kind of control of the fleet allows for a more rational use of the limited resources that the company has available to it. However, external environmental forces affect organization of work on the vessel as well.

THE IMPACT OF EXTERNAL ENVIRONMENTAL FORCES

During the late 1970s and early 1980s a number of crucial changes took place in the industry. In 1977 Canada declared a two-hundred-mile exclusive economic zone around its shores, and had to devise ways of managing the additional fish stocks. At the same time, with a crisis in the market-place, the fast food industry required a better quality of fish than Canadian enterprises had been supplying. More-over, many of these secondary processors moved away from fish to chicken, an easier-to-handle product with a greater demand. This resulted in a loss of many of the markets for Canadian fresh fish and a financial crisis for most companies. The major fish companies in Atlantic Canada could not continue to compete without a major overhaul of their industry – a costly undertaking for an industry already in serious financial trouble. The government's major reorga-nization and refinancing of the deep sea fishery after *Navigating*

Troubled Waters (the Kirby Report), coupled with the introduction of the enterprise quota, drastically changed the nature of the fishery. In the interests of efficiency and financial responsibility, companies introduced scientific or rational management procedures.[13]

After the establishment of the two-hundred-mile limit the state, in cooperation with company officials and fishers, developed new management policies for controlling the resources of the deep sea fishery. One of these policies, the *enterprise allocation,* allowed individual companies to receive their own separate quotas, thus privatizing the resource. Under the old allocation system the industrial sector of the fishery received the quota and individual companies competed at sea for their share of the permitted catch. The new system allowed for the rational management[14] of the resource by eliminating competition at sea for total volumes of fish. The change of emphasis from quantity to quality of fish caught led to a more rational management of primary processing and to more value-added processing.[15]

The enterprise allocation in the deep sea fishery in Eastern Canada now applies to the groundfish, deep sea scallops, lobster, shrimp, herring, and surfclam fisheries. In 1982 deep sea trawler companies negotiated the first enterprise allocations for vessels over one hundred feet, in response to licence limitations and a rapid growth in the midshore/coastal fleet. The enterprise allocation met the needs of both the government and the companies. The government wanted to control the fishery more closely, and the companies, in order to respond better to market demands, wanted to introduce scientific management procedures and improve fish quality. Both company and government officials believed that the enterprise allocation would remove some of the uncertainty in the fishery by stabilizing the size of the catch, the length of the trip, the quality of fish, and the wages of workers. In addition they believed that increased control of the fishery would increase safety. Fishing companies welcomed the changes in the light of these anticipated benefits.

Intra-industry negotiations set initial allocations, except for the lucrative northern cod fishery located off Newfoundland and Labrador, where the Department of Fisheries and Oceans (DFO)[16] acted as arbitrator.[17] They based these initial allocations on landings over the previous ten years, with the average of the first seven equally weighted to the last three years' landings. When the DFO withdrew from arbitrating the voluntary program in 1983, the industry continued to observe the agreed-upon allocations. In 1984, the DFO formalized the program as a five-year experiment. Since then enterprise allocation has been voluntarily expanded to include the deep sea scallop, lobster, shrimp, herring, and surfclam fisheries. Since 1984 nearly one hundred companies – sixteen large companies with multipurpose fleets of vessels over one hundred feet, and approximately eighty smaller companies with vessels in the sixty-five to one-hundred-foot range – have been included in the program.[18] Discussions have now moved on to individual transferable quotas, similar to enterprise allocations, for vessels under sixty-five feet.

The advantages of the enterprise allocation system over the previous quota system relate to the timing and location of competition in the fisheries. Under the old system each sector of the fishery – whether deep sea or coastal – acquired a quota for catches of a species in a specific area during a particular period. The companies competed for the resource on the high seas, catching as many fish as possible, giving lower priority to the quality of the fish. Companies required a continuous supply of fish to keep their plants operating, so the industry-wide restriction on the amount of fish to be caught in a given area during a specific period produced an emphasis on quantity, not quality.

Under the enterprise allocation system, each vertically integrated company has its own allocation for each species of fish in a given area over a specific period. Competition for the allocation takes place on land, in boardrooms, well before the ships head to sea. This procedure allows more rational planning for the use of company facilities. Companies can respond to the needs of the markets and plan their plants' operations accordingly. Similarly the plants' needs can be met by sending vessels to a particular area for a specific species of fish. With no need to maintain a large inventory of processed or unprocessed fish, fish can be harvested "just in time" to meet the plants' demands. With a limited quantity of fish, value must be added either through improved quality, or through secondary processing (for example, frozen complete dinners, breading of fillets). Market conditions and the allocation system make an emphasis on a better quality product both attractive and possible.

Enterprise allocation cannot be evaluated in isolation, and changes in the allocation system must be seen as part of a business strategy. Harvesting of each species takes place when the company determines it to be most advantageous. A brief description of a fishing trip illustrates how this environment affects working conditions.

A TYPICAL TRIP

Before one of the fishing trips I joined,[19] the vessel and crew had sailed from their home port in Nova Scotia to the Grand Banks, off Newfoundland. There they fished for groundfish, and after eight days, the company routed them to a southern Newfoundland port where they unloaded their catch. It would take the crew a minimum of nine hours' travelling time to reach their home port in Nova Scotia. Under the crew's collective agreement the company could request the crew to take their leave of forty-eight hours in the Newfoundland port and then sail. In return they would receive an additional twenty-four hours on their next leave.

Instead of returning home to Nova Scotia, the crew spent forty-eight hours in Newfoundland, staying at the local hotel at the company's expense. The company then sent them to the Davis Strait area off Greenland to catch turbot. There they fished for another eight days, and landed at a northern Newfoundland port. Under their collective agreement, they had to return home after their second voyage. The trip back to their home port required an eight- to nine-hour journey by plane and taxi.

I joined the crew after their seventy-two-hour leave in Nova Scotia. We set out for Newfoundland at three in the afternoon and arrived that evening just before eleven. Although we had planned to leave for the Davis Strait that evening, the vessel did not sail until the next morning. On the second day, just off St. Anthony at the northern tip of Newfoundland, the company told the captain to change course, and head toward the Gulf of St. Lawrence off Port aux Basques, Newfoundland, and to fish for ocean perch (redfish) there.

Steaming to the Gulf of St. Lawrence from St. Anthony took one and a half days. Since fishing for ocean perch requires a midwater trawl rather than a ground trawl, the crew spent this time changing, checking, and repairing the nets. Although highly adaptable, the midwater trawl can be difficult to operate. The tactic in midwater trawling consists of locating the fish by coordinating the data from the fish-finder and the transducer. The fish-finder aboard the vessel, and the transducer, which is sewn into the net itself, indicate the terrain of the seabed and the depth of the schools of fish. As the fish move up and down in the water the vessel follows them, dragging the trawl behind, catching the fish in the net as the trawler overtakes the school. During this procedure the operator must also be aware of the sea bed's terrain so that the net does not get snagged. (Figure 3.5 illustrates midwater trawling as well as the use of sonar in locating fish and determining fishing methods.)

These last-minute changes in operations posed additional problems in the running of the ship. The regular captain had taken this trip off to go on a family vacation. Only the acting captain, who usually sailed as mate, had any experience in running a midwater trawl. The acting mate, usually the bosun, had little experience on the bridge. The acting bosun, an older worker in his fifties, usually sailed as a cook, although he held a Third Class ticket and had previously sailed as bosun. Although on paper the officers held adequate credentials, in reality the bridge lacked staff with experience in midwater trawling. This meant that the acting captain seldom left the bridge; he never left during the hauling back or shooting away of the net. As the trip progressed, stress and fatigue took its toll on him. He hollered at the crew, and in hauling back the trawl too quickly they lost a door. His lack of attention to the sea terrain resulted in a ripped net, the loss of the catch, and a further hour of lost fishing time while the crew repaired the net.

As the hold began to fill, it became apparent that the volume of ice available could not adequately store the amount of redfish specified by the company – redfish require more ice than the turbot they had originally planned to catch. This created a dilemma for the acting captain: should he catch more fish, which would result in a product of lower quality, or should he catch fewer fish and maintain better quality? He had to choose between the economic well-being of the crew or the company. The more fish caught, even at lower quality, the better the crew's wage. The captain decided on quantity, thus increasing the crew's pay. As the hold filled, the captain informed the company.

At this point, the acting captain did not know where he would be instructed to land the vessel. Two ports, one in southern Newfoundland and one in Cape Breton, were possible landing points. The crew showed their frustration by making bets on the possible receiving plants and time of landing. Before leaving Nova Scotia, the company had told the crew to assume a two-trip cycle out of the northern Newfoundland port. The crew had advised their families of this and many wives had assumed that their husbands would not be home for three or four weeks and had made no arrangements for an early return. But on the morning of the day we landed, the company finally told us where we would dock – Cape Breton. We landed at ten o'clock in the evening and got into taxis to travel to our home port. Although we reached home at about four o'clock in the morning, most of the men's wives and young children greeted them in the parking lot, where many had been waiting since early morning.

CHAPTER FOUR

The Social Context: Wives and Families

The deep sea fisheries feature a rhythm of ten days of work at sea punctuated by forty-eight hours of leisure on shore.[1] This chapter examines the extraordinary pressures on deep sea fishers' households, and how a woman's life can be dominated by the nature of her husband's work in the deep sea fisheries. This "open letter" published in *National Fisherman,* a magazine widely read in North American fishing communities, speaks to these problems:

To the Editor:
There's more to the fishing industry than your magazine prints. These tough guys are real people with real families, and more than a few have serious drinking, drug and emotional problems.

Whether fishing produces these flaws or draws them to it, I don't know. It's not the time and distance that destroys these guys' families; it's the hard-core attitude they get when they hit land. They are very loving on the sideband [ship-to-shore radio].

If you choose to print this [open letter], please don't print my name or address:

We don't think you know how much we appreciate the hard and exhausting and dangerous work you do, or that we understand what your blood, sweat and tears produces. We're thankful for our home, our treasures and the food we eat.

We don't know how to reach you anymore. Our lives are as vast as the sea from which you fish. We want to love you and respect you. We keep coming up with a broker.

Every time you leave, you take other people's souls with you. We sit looking at the full moon, wondering how and where you are. The kids draw pictures of your boat and tell everyone with such pride in their eyes, "My daddy's a fisherman."

We spend our time taking care of your home, so it will be clean and warm when your ship comes in. We realize it's not an easy transition from your world to ours, but if you will trust us and let us, we will help you make the adjustment.

After all, we're hooked on you, and we are the best catch you ever hauled in. (Anonymous 1988)

All deep sea fishers' wives face a common dilemma: those factors that facilitate their adaptation to the role of single parent/person may conflict with their fulfilling the role of wife. During her husband's absences, a woman takes on the responsibility of single parenthood/ personhood, and then must reassume the role of partner on his return. Deep sea fishers face a parallel dilemma: their adaptation to life at sea may conflict with their role as husband/father and their integration into the shore-based community. These men's allegiance wavers between their family and their fellow crew members – their family at sea.

The responses to these dilemmas differ, and stem from the variations in the structure of the fisheries, the history of the fishing communities, and the social origins and life-cycle stage of the fishing households. Historically, fishing families exercised a strict gender division of labour. Men caught and gutted the fish, while women and children processed the fish on shore (Antler 1981; Faris 1966). In the household, the locus of activity, labour separated into "men's" and "women's" work. The wife's work complemented the work her husband did. The industrialization of the fishery eroded the complementarity of men's and women's work and removed it entirely from its former context of "family" work based in the household. The separation of men's and women's work facilitated the industrialization of the fishery. Men still catch the fish, but they do so on company-owned and -operated vessels. Women continue to process the fish, but in industrialized plants.[2] Women run the households with little or no aid from their spouses, who support the households on their wages.

THE COMMON PROBLEMS

The nature of the deep sea fishery dominates the organization of family life in the fishers' households. Men work on the deep sea trawlers and scallopers, women maintain the household and take care of the children. Defined and confined by the constraints placed on women by their husband's job, women's work varies depending on the frequency and duration of husband-absence.

The conflicting demands of work and family responsibilities aggravate tensions between spouses. These tensions occur in all households where women depend on men as the primary wage earners (Luxton 1980; Luxton and Rosenberg 1986; Oakley 1974). The prolonged absences at sea, high stress, and physical risks exaggerate these problems and distinguish fishers' families from other households. These characteristics also make it appropriate to compare families of army and navy personnel with families of fishers.

Prolonged periods of husband-absence generate stress for both the husband and the wife in any setting. In a study of the Canadian Forces, Truscott and Flemming (1986, 21) found that "those who reported the most unaccompanied tours in the past five years, the most separations of more than one week duration during the past year, and the most total months away from the family in the past year were also most likely to report high occupational stress."

Besides higher occupational stress, these participants also reported a perception of greater conflict between job and family activities (Truscott and Flemming 1986, 13). In a separate article on the same study, Popoff and Truscott (1986, 8) document that the spouses of service members experienced symptoms of anxiety and depression more frequently than the service members themselves, suggesting that separation and reunion create particular difficulties for the spouse who remains at home. The nature of these difficulties depends on the stage of the temporal sequence: separation, homecoming, and life together.

THE SEPARATION

While deep sea fishers fish, their wives function as single persons/ parents. They assume the day-to-day responsibility of running a household, managing on their own, or with the help of family and friends. They must develop their own social networks and lives. In speaking of American Navy spouses, Decker (1978, 114) comments that "What may be perceived as an opportunity at one point, may be a stress or crisis at another." This applies to the separation period for deep sea fishers' wives as well, particularly to issues such as independence among the wives. The lack of a mate to depend on undoubtedly causes stress for some, while for others it may lead to a confident and independent sense of control. In a study of the wives of American prisoners of war, McCubbin and Dahl (1976, 114) state: "The waiting wife, functioning as head of the household, often matures, develops greater independence and self-confidence, and provides a life style for the family in the absence of a husband or father."

Uncertainty pervades a fisher's wife's life during the separation period. When her husband leaves on a fishing trip, she does not know when he will return, nor does she usually know where he will fish. He has simply gone on a trip. By listening to the CB or ship-to-shore transmissions (boats can have either), and by talking to other wives of crew members, she may get a sense of how the fishing progresses. She knows he will not be coming home until he has a "trip" (a full hold) or the fish begin to spoil. She "expects him when she sees him." Unless he "takes off" a trip for a specific event, she can never plan for him to be home for anything: the servicing of the car, family outings, school activities, the birth of a child, a miscarriage, or the death of a parent. Even when tragedy strikes he cannot come home.

The stress this situation places on individuals and their relationships is illustrated by the following example:

Then I got pregnant a couple of months later with my second little girl. She was born premature ... and Charlie was out [at sea]. They had to call him ship-to-shore ... I didn't know how he got the message. I just wanted him there and he wasn't. I had my father of course and Dad told him the little girl had died. This is how he got the message ... that I was fine but the baby died. He couldn't understand it. He thought it was the older girl. You see you can't come in. (Wives Interviews)

This lack of shared experience can undermine the relationship and pull the couple further apart. Each individual experiences the same event in quite different ways:

So I guess Charlie went through a rough time out there. By the time he got in, the little girl was buried. He didn't get a chance to see her. It bothered me for a while ... I carried her, to me she was real, but it was so much harder for him to feel that same kind of loss because he didn't even get to see her. He still talks about it ... that he wasn't there to help with it. (Wives Interviews)

The characteristics of her husband's work dominate the wife's work and leisure. The fear that her husband may return from sea maimed – or not return at all – creates a constant worry. As one woman put it:

At first I was real frightened. If there was a storm I'd be crying, if they were delayed, a day late or I'd hear they had ice, I was sure he was hurt. But after a short time I accepted it. I had to otherwise I would be 'brokeup' [upset] all the time. Still it's always there at the back of my mind, I just try not to think about it. (Wives Interviews)

This same woman's description of her best friend illustrates how the fear can be more pervasive.

She's my best friend but we're altogether different. I'll try to calm her down but she just paces, paces, and says 'We're never going to see them again, they're not coming back this time'... She worries about everything. And she has one of those radio sets – a lot of women have them, but I can't imagine why they would want them ... the ship-to-shore ones, they can hear the plant talking to the ships, or to other boats, and you know everything then. You know if they've got problems with the winds, or in the engine-room. And you are better off not knowing ... So the women hear that there is something wrong with the main engine ... and they'll get all in a panic ... they'll picture the boat sinking, but you know if they're staying out there, if no one's sent to get them they must be okay. (Wives Interviews)

Constant concern about the present and the fear of an uncertain future characterizes these women's daily lives.[3]

Finances are equally uncertain. Deep sea fishers receive as earnings a share based on the quantity and quality of fish caught. The drive for quality and scientific management practices have made the vacillation in fishers' earnings less extreme, but substantial variation in income, still occurs depending on the species of fish caught and their current market value. Hence, fishers' wages fluctuate from trip to trip, making financial planning difficult. Not until the catch lands does the fisher know the amount of his pay. As one woman explained:

Before unions, those men had nothing. They'd come in, they'd land, they wouldn't see five cents until it was time to go out. They'd get their cheque the day they were leaving, so we handled the money. I would say most women still do. But since the union ... the men get an advance when they land, one hundred or two hundred dollars. (Wives Interviews)

Sometimes, though, they have no money to draw on.

Now last trip was a "broker." They settled for seadays ... You know, you're given so much a day ... Once at Christmas time his cheque was fifteen dollars and some cents ... But if that's your only income, I don't know what you'd do. (Wives Interviews)

In difficult financial times, the wife may approach the company, local stores, or family for an advance against future income, but short-term credit does not help the long-term budgetary problem of not knowing how much money will be available in a given month for family/household expenses. Many women have never handled the finances before and this adds to their anxiety.

The handling of the money falls to us women ... If the poor woman couldn't handle it, if it meant for her sanity (if she had children) getting to bingo two or three times a week, that and a baby sitter, well a lot ... ran into difficulty ... Because the women had never been taught to handle money ... never observed someone else do it either. (Wives Interviews)

Wives adopt various strategies to meet financial demands, such as living with parents, living in a trailer, building a house bit by bit, or seeking full- or part-time employment.

Paid work helps stabilize the household income and increases wives' independence by giving them money of their own. However, getting and keeping a job presents additional problems with household management. Jobs in industry usually involve shift work, with eight-hour shifts for five days or twelve-hour shifts for four days with two days off. But such a job schedule does not mesh with the deep sea fishing cycle. Connolly and MacDonald (1985, 416) argue that this factor accounts for the "persistently lower labour force participation rates" of fishers' spouses compared to fish-plant workers' wives. Few jobs have, or will allow, the flexibility needed for the type of short notice that fishers' wives must operate with. Part-time work, when available, furnishes a partial solution but provides only some of the benefits of employment. It does give a fisher's wife a greater probability, although far from guaranteed, of being at home when he expects her to be. Often a wife calls in sick when the husband has shore leave. Thus she places the independence and other benefits of having a paid job at risk so that she can "be there for him."

Wives are more likely to seek a paid job in the younger fishing households. They need money to get started and have no savings to draw on if a run of poor catches occurs. Once they have children, their financial problems worsen, especially when they need to pay for child care. As one woman put it:

Then the finances. There were times, especially starting out when we were new, so little money and you didn't know, you'd rob Peter to pay Paul, you'd just love to be able to say to him "Take it, you look after these" especially after we decided we'd be building our own home. Of course I was lucky I'd started back to work. That was another thing when I first got married, that people here expected that I was going to stay home and bring up children. That was fine by me until you had one and you realized you can't make it on just one income, not if you want a home ... Money can be a real problem. Some of those women alone that aren't working, I don't know financially how they did it. (Wives Interviews)

In many fishing communities formal day care does not exist; when it does, it usually provides crowded, expensive, or seasonal facilities. Many women feel that putting their children in day care indicates that they neglect them. When day care centres provide inadequate care due to underfunding, understaffing, or lack of equipment, mothers use them relucantly. They channel their energies into informal, individual child care solutions that at least permit them to think of themselves as good mothers. This perpetuates the problem of too few formal day care facilities geared to fishing families.

For those families with relatives nearby, the employment of the wife may continue after the birth of the first child, but ordinarily not after the arrival of the second. For help in child care, most women rely on extended family, usually mothers and sisters and in-laws, or on wives of other crew members who live in the area.

It's a new experience, especially your first child, and you're alone so much. Some of them wouldn't feel it as much as I did because I wasn't from here. Like Jane and Susan, [two sisters whose husbands are on the boats], they have their mom there, and their sister Mary ... they made use of their extended family here ... That's probably common in fishing. (Wives Interviews)

This reliance by the wife on kin and friendship-based child care creates a double dependency on husband, parents, and friends, which has both emotional and social costs and leaves her vulnerable to social control. For example, one woman talked about how her mother-in-law would curtail her going out at nights by refusing to babysit for those occasions that the mother-in-law deemed inappropriate. Even when the wife's mother babysits the first child, there may be reluctance – on both their parts – to continue this arrangement after the birth of a second child. In every situation the wife's child care responsibilities, the family's ability to pay for day care, or the ability to put up with the difficulties inherent in depending on informal day care, limit the wife's ability to work.

During the separation period wives develop a social life discrete from their husbands'. Frequently the wives of crew members form a social unit. This creates its own special difficulties, often severe ones, particularly if they frequent restaurants and bars. The more conservative members of the community label them "lounge lizards." Husbands may get reports of their wives' activities, and jealousy and fears of infidelity abound. The misbehaviour of fishers' wives can be viewed as an indication of their husbands' inability to control them, and this perception may itself lead to conflict between spouses. For the wives, socializing together may be a form of joint rebellion, where each supports the other in dealing, after a fashion, with the common problem of being a deep sea fisher's wife. The wives feel that this helps them cope with their particular problems better than contact with kin does. Yet they pay the price of social disapproval for this freedom.

The port from which a fisher sails affects the choice of the community where his family will live. This choice of residence profoundly influences his wife's way of life, especially during his absence at sea. Women with family or in-laws in the area of the port have ready access to friendship and support networks. In the case of migrant fishers who do not live in or nearby the port from which they sail, their families may follow them or remain in their home community, depending on the distance from the deep sea port. A distant port location can generate many problems. If the wives of migrant fishers have a long distance to travel to the deep sea vessel's port, they tend to see their husbands less often. In Lunenburg, a number of fishers have families living in Newfoundland, whom they see very seldom except when they have Christmas vacation, they take off a trip, the vessel undergoes refit, or the family comes to see them. These long-distance relationships put additional stress on their families, although the wives usually have a strong support network in the home communities to meet the problems of everyday living.

A more frequent option for migrant fishers, especially for those having seniority and some job security, involves having their wives and families move to the deep sea port. This solution allows the wife and family more time with the husband/father, but these households arrive in a community as strangers isolated from their extended family and friends. Most of these women develop support networks with other migrant women or with wives of other men who work on the same vessel. Commonly, they also rely on community resources – church associations, clergy, social service workers, doctors and other medical support groups – to a greater extent than nonmigrant fishermen's wives.

Social contacts may also be further limited by community expectations. Only certain social events, such as social outings with the immediate and extended family, bingo, church socials, and "home and school" functions, usually in the company of other women or family members, may be seen as appropriate for the wife of an absent fisherman.[4]

Of course, any outings may be impossible for many of the migrant fishing households. As one migrant woman noted,

I don't know how we coped that first year and a half, plus we had no car. The baby on top of it. At three o'clock in the morning, Janet'd be crying and I'd be crying ... Because I just had no one. And my mom was in Newfoundland, I was the eldest of nine, seven of us living, and mom still had children at home. Mom couldn't drop everything and come. These women here had that advantage. (Wives Interviews)

No matter how much emotional support these women receive, they must shoulder the responsibility of raising their children alone. Migrant women must make the day-to-day decisions concerning child care, be it formal day care centres or friends in the community, and must be the principal disciplinarians. This can be particularly difficult with small children. But as the following example illustrates, these responsibilities continue into the teenage years.

I don't know, I've only had girls, but I know the teenage years are the hardest ... what to say "yes" and "no" to, and sticking to it, I found extremely difficult ... Some of those decisions you have to make then, not when Daddy comes home. It's too late you know, if it's a dance on Friday night. You can't say, "Well, your father'll be in next Monday, we'll talk about it." The dance is over. So those things are very hard ... And then, "I know if my father was here he'd say yes," and maybe he would have, but I didn't want that kind of responsibility alone even if it was just a dance ... (Wives Interviews)

The choice of residence affects the economic and social opportunities for fishers' wives. The port facilities for the deep sea fleet lie beside modern industrial processing plants, located in a few large communities with labour available for both plant and vessel needs. Other businesses in the community supply services to the fishing industries and their employees, but they also supply similar services to related industries. As a result the community has a wide range of economic opportunities available to women, as well as major social service facilities such as transition houses, detox centres, and day care. Opportunities for leisure activities abound, and the more cosmopolitan nature of the community makes a wider range of activities socially acceptable.

In smaller communities based primarily on fishing, few economic opportunities for women exist outside of the fishery, except in the local restaurant, school, or hospital. Women may have the support of family and friends during their husbands' absences at sea, but they have few opportunities to get out except to go to bingo, church-sponsored events, and family gatherings. Few social services besides the clergy and the local doctor exist. When a relationship turns sour these women find it difficult to get professional or impartial help.

The choice of community also affects the fisher's social contacts. His work makes it difficult to maintain or develop social contacts outside of his shipmates. If his family continues to live in his home community he must travel, in some cases for many hours, to reach home. But when he gets home he has his extended family and friends. If his family takes up residence in the deep sea port, he has the convenience of having his family right there, but he has no social network of his own except for his fellow crew members. This lack of social and community interaction increases the fisher's sense of alienation, and increases the likelihood that he will spend his shore leave almost exclusively with his shipmates. This situation puts additional pressure on wives, not only to develop their own social network but also to develop a "couples" network. The easiest answer to this problem involves developing social networks with other crewmen's wives.

THE HOMECOMING

The homecoming begins just before the vessel arrives. Wives phone the company's "hot line" daily to listen to a prerecorded message giving each boat's progress and return date. The boats usually dock between midnight and six in the morning. Despite the uncertainty of life alone, men expect their wives to drop everything and to be available for them on their return.[5] They insist that their wives either pick them up or have their cars waiting near the dock. As one fisher explained,

When I come up over the dock, she better have the car turned around, the driver's door open, the trunk open and be sitting on the passenger's side waiting for me ... My car not being there really spites me ... I just hate waiting to go home. I only have forty-eight hours. (Well Workers Interviews, Deckhand)

In many cases, this means bundling up young children, putting them in the back seat of the car, and driving to and parking on the wharf with a thermos of hot coffee and the sleeping children. There the women wait for the vessel to appear on the horizon and to steam slowly to the plant's wharf, passing the time by chatting with the other waiting wives. The air bursts with excitement. One woman describes her wait on the dock:

You know, they go out for ten days and you get all excited, especially now that the kids are bigger, because you hear the boat blowing and you've got to run right away. You stand on the wharf for half an hour or so. But I think it's a real plus ... I can't wait until he gets home. (Wives Interviews)

Although it is the only time for the renewal of the husband/wife relationship, both partners find the homecoming stressful because of their different expectations. These expectations and the disputes that arise from them do not differ in context from those that take place at the end of the day in homes where men return daily (Luxton 1980; Rubin 1976; Oakley 1974; Luxton and Rosenberg 1986), but the prolonged absence of the husband heightens the importance of these concerns. The wife looks forward to her husband's return to give her companionship, a respite from having sole responsibility for the finances and family, and the brief resumption of a social life as part of a couple. On the other hand, when the weary and tired fisher finally leaves the vessel, he looks forward to relaxation, time with children, and a rest from the pressures of work at sea. The mismatched expectations must be played out. Compromises must be made.

The wife anticipates socializing with people other than family and women friends. Her husband wants to live ten days in two; he wants to see his buddies, participate in recreational activities, and play with his kids. Fishers need to rest and relax after coming home exhausted. In some cases, they may have landed in another province, had a two- to five-hour drive to the airport, a two- or three-hour flight, and then another drive to their home port. In other cases, men have chosen to live in smaller fishing communities where their families have lived for many generations, and commute to the port where the company moors its vessels. In either case men routinely return home exhausted, not only from their work but also from the subsequent long journey home. Sometimes, wives will have to meet a husband who is not only fatigued, but has been drinking on the trip home.

Although the wife runs the household, she depends on her husband's wages to pay the bills, and she has no way of knowing what he has been paid until he gets home. It may be a good or a bad catch, and she must budget without knowing how much she has to spend. The tension associated with uncertain finances may lead to domestic disputes where emotional responsibilities get confused with monetary responsibilities. Sometimes the wife becomes characterized as more anxious about the pay-cheque than her husband – the woman on the dock who shouts to her husband still on board the vessel, "How much did you make and when are you leaving?" He becomes either the reckless spender who does not care for his family's well-being and who leaves his wife with unpaid household bills, or the spouse who feels he needs only to provide monetary support and neglects his wife's emotional needs.

Since the wife has been the responsible adult in the family during her husband's absence, she wants a break from parenting and wants him to take on some of the child care. She feels the children need and want to interact with their father. As well children may need discipline, arbitration, and support.

Sometimes ... oh if only he were here, to let them storm at him or go in their rooms pouting, for him to see it. Because his girls can do no wrong ... when he's home, and it's true, they're not bad kids at all, but he ... didn't see all the little things that brought them where they are, and he didn't have to listen to some nice crying. They came out and shouted "You're just mean and I know Daddy would have said yes." But those are things some parents go through even if their husbands are home. I am so glad when he comes home then he can deal with it for a while. (Wives Interviews)

Disputes also arise over disciplining the children. The husband may, unintentionally or not, contradict the decisions made by his wife during his absence. He may hand out more or less lenient punishments or question the need for punishment at all. This undermining of his wife's authority weakens not only the couple's relationship as parents but the relationship between mother and children.

He wants to get to know his children. He feels like a stranger in his own home, and the family appears to function without him. His young children may not recognize him and the older ones ignore him. He may attempt to woo his family's affection with expensive toys, such as four-wheel recreational vehicles for his twelve-year-old son or a ghetto-blaster for the pre-teen daughter. In some cases, as reflected in the negative commentary – so common among fishers at sea – about women's complaints, the husband begins to resent his wife and sees her as the lucky one, staying ashore and enjoying family life. He wonders why he works so hard and long. Many fishers cite lack of participation in family life as one of the most common complaints about working the deep sea, and as one of the most common reasons for leaving the deep sea fishery.

Homecoming encompasses more than mismatched expectations; the wife is often concerned about her husband's reaction to progressive changes in her behaviour. McCubbin and Dahl (1976, 139) had this to say about the concerns of the wives of returning American prisoners of war:

Realistic appraisal of the wives' concerns and apprehensions about repatriation also suggested that the anticipation of reunion posed a threat to one or more of the rewards that the separation had provided, e.g. the opportunity to assume greater freedom, an independent income with the latitude to determine its use, and the avoidance of any confrontation with their husbands about the manner in which the wives conducted themselves during the husbands' absence.

Nearly half the wives of these repatriated Americans primarily feared their husbands' reaction to their increased independence.

The homecoming often involves a struggle between husband and wife. The same "open letter" quoted at the start of this chapter went on to summarise common complaints by wives:

How to lose a family:

- After a 21-day fishing trip, spend the first night home with the guys celebrating your catch.
- When you finally do make it home, don't forget to accuse your spouse of messing around with every guy left on land while you were gone. You could even slap her around, just to make sure she won't do it next time you are gone.
- Criticize the kids for the way they mowed the yard and painted the garage. Why should you thank them? You pay the bills.
- Get good and drunk the night before your only day off. That way you'll sleep soundly the next day, and you won't have to worry about being amorous.
- When you finally do wake up, don't play with the kids; they will be much more fun when they're older. Besides you haven't watched a game on TV with your buddies lately.
- Send your wife to the store on a beer run, a cigarette run and a pizza run. After all, what's she there for anyway?
- Spend all your money in the bar before you leave. Then you won't have to give any to the rug rats for summer camp. (Anonymous 1988)

Disputes arise from fears of infidelity, sexual jealousies, finances, discipline of children, and the husband's apparent lack of interest in children or household concerns. Domestic violence often occurs, usually within the first twenty-four hours of the husband's coming home. In many cases alcohol contributes to the situation. Most women stay in the marriage and try to make things work. When their husbands' misbehaviour (abuse, drinking, fighting) has gone too far, wives call it quits and return to their families, live on their own, or start a new relationship – preferably with a landsman. The first and last choices appear to be the most common.

In the major deep sea fishing centres in Nova Scotia, a high proportion of deep sea fishers' wives and family members receive counselling at local transition houses and abuse centres. In some areas over sixty-five per cent of clients come from fishing households.[6] Although these high numbers may simply reflect the demography of the community, the characteristics of this fishery enhance the probability of spousal and family abuse: a matri-focal family life, a high-stress industrialized workplace, and little time to work out differences. Lenore Walker (1984) argues that risk factors for battering occur in families where insecurity, jealousy, alcohol, and substance abuse aggravate conflicts for men with traditional values concerning women's roles. The long separation and short reunion cycle clearly threaten the husband–wife relationship. The erosion of autonomy, and the high-stress, high-risk environment of the deep sea fishery increase the anxiety and erode the self-esteem of fishers regarding their family relationships. They work hard to provide for their families. But when they return from the sea, they enter an unfamiliar world. They have even less control there than at sea. The couple must develop a way of coping with this frustration or the marriage will fail.

LIFE TOGETHER

Life together usually consists of the few hours between fishing trips. But a few exceptions exist: the summer refit (traditionally the time for the annual family holiday), the winter layover, time off for accidents and injuries, and layoffs. Some companies tie up and refit all of their vessels at the same time. Other companies refit their vessels in sequence; everyone on the boat knows the order in which the vessels go for refit and where their vessel comes in the order. In those cases vacation plans cannot be confirmed until close to the date. The annual refit provides the only time the whole family can go away together, and families look forward to this event with great excitement. The smaller vessels (sixty-five to one hundred feet) also have a winter layover – depending on the weather – from late November/December to March/April. Many men take advantage of this layover to work part time, go lobstering, fish, bring in the wood, or go back to school for further training.

Living together requires readjustment for both spouses. He must adjust to his family, and leave his life at sea, the crew – his other family – and its male world, behind. He enters a matri-focal household where he may feel like the outsider. And his wife must incorporate his needs into her daily routine:

When Paul is gone, I go to work, I come home, I look after the kids, I do a load of wash, I can iron if I feel like it, I go to bed any time I feel like it. But if he's home, he upsets my routine. Now that bugs me, someone else it wouldn't bug. Because Paul will say "Oh leave that there and come sit down with me" ... There's nights that I'd ten times sooner be doing ironing and he'll say "Come sit down, come watch this with me," so you'll leave it there and sometimes you feel they're underfoot. (Wives Interviews)

When he returns home, his needs dominate: his laundry must be done, his favourite meals cooked, his entertainment needs met. Husbands give little or no recognition to what their wives give up in order to accommodate them or their needs.

While she balances her daily routine with his presence, he has to cope with an unfamiliar shore routine and separation from the male world of shipboard life. Often a fisher finds he has little in common with nonfishers and has no role in the community other than husband.[7] He wants the respect of the community and his family. Many fishers trade off being at home for financial well-being. The prevalence of expensive consumer goods such as fancy new cars, modern houses, all-terrain vehicles and other leisure items, characteristic of deep sea fishing communities, indicates his success and proves that he is a good provider, husband, and father. As in other industrialized settings, conspicuous consumption "proves" that one has succeeded. This outward evidence of success validates their choice of working in the deep sea fishery rather than in other employment.

Although the acquisition of ostentatious goods may originate in the objective reward structure of deep sea fishing, additional factors reinforce conspicuous consumption. From the husband's point of view, providing the convenience and the status of consumer goods in addition to the basic income is an enticingly easy way of compensating for being an absent father and husband, of discharging familial obligations, and of excusing himself from extensive domestic involvement while on shore. Some wives support their husbands' behaviour because it consoles them and rationalizes their husbands' absence. If this interpretation proves correct, the level of conspicuous consumption would predict the extent to which the husband expects the wife to serve his needs, as well as the likelihood of his spending his time at home either with his shipmates or in leisure activities with his older male children. Friendships with crew who give conspicuous consumption high priority would also provide role models and social pressures to follow the same lifestyle. This priority would be particularly disappointing and stressful in those households where the wife has structured her life to seek rewards through marital companionship during his time ashore.

In response to the demands of their dual lives, fishers often fall into one of three types of behaviour – shipmate, family man, and honeymooner. The "shipmate" fisher spends as much of his time as possible with other crew members, going home to change his clothes, sleep, and eat. He spends little or no time with his immediate family. Although his wife may come along with him to parties or dances, most of his leisure activities consist of male pursuits such as hunting and fishing, hockey and football, card-playing and drinking. Usually, the single men of the crew and migrants who live in the deep sea port direct this group.

The fisher with the second pattern of behaviour, the "family man," typically spends as much of his shore time as possible with his immediate and extended family and his close friends. For example, he works around the house, visits his parents, goes out in the woods with his kids, or goes out with his wife. He would seldom be involved in activities associated with the crew except for special functions such as a wedding reception for a crew member. Usually he lives in his home port.

In between these extremes lie those who forge a balance between commitments to family and crew. For example, the husband might play touch football during the afternoon with the crew and in the evening take his wife out to dinner and the movies. Women talk about "taming" their husbands – changing them from the "shipmate" pattern to the "family man" pattern.

The "honeymooner," an intermediate pattern of behaviour, encompasses fishers who spend all their time with their wives, doing things the couple want to do with each other. If they have children, the wife has usually taken them to a relative or friend who has agreed to take care of them for the duration of the husband's shore leave. Although this pattern may maintain and strengthen the husband–wife commitment it prevents father–child bonding.

WHAT MAKES THE DIFFERENCE?

Whether or not a couple can cope with the pressures associated with this way of life depends on a number of factors. The availability of family and friends for social support and child care relates directly to household residency. In the industrialized deep sea fishing communities of the Lunenburg/Bridgewater area, where a sizeable proportion of crew members come from Newfoundland, wives with no family in the community find it particularly difficult. Residential patterns reinforce their isolation, since they tend to live on the periphery of the communities. Under rationalization, fishing companies with plants in different locations dispatch crews and vessels to these ports, often making it impossible for fishers to come home between trips. As well, individual fishers may take advantage of promotion opportunities by changing home ports; in some cases their new "home" port may be in a different province than their home community.

The cyclical nature of the fishery means that employment opportunities fluctuate depending on the fish stocks. When employment demands in the deep sea fishery rise, the coastal fishery, the oil industry, and the onshore industrial companies like Michelin Tire compete with it for labour. In this environment the number of "commuter crews" will increase, as will the concomitant social isolation of their families. When employment demands in the deep sea fishery fall, fishers must seek alternative employment and many return to their home communities.

Another factor in the durability of a fisher's marriage includes whether the household comes from a schooner (deep sea) or a coastal fishing tradition. Schooner fishing and deep sea fishing impact on crews' wives in similar ways. The fisher's wife who comes from a schooner or a deep sea fishing family has probably been brought up with an appropriate set of expectations, and come to her own marriage "equipped" with a repertoire of effective coping skills. In contrast, those from coastal fishing families likely hold expectations inappropriate to deep sea fishing families. As one deep sea fisher's wife said, "My father was an inshore fisherman so I thought I knew what fishing was all about. But that type of fishing – dragger fishing – why, it was all different. I didn't know anything about it" (Wives Interviews).

A third factor is the extent to which the couple communicate. Here is one fisher's description of how he and his second wife have worked it out: "In a little over three years since we've known each other, we haven't had an argument yet. We have our differences but we sit down and talk about them. Everything is going just fine" (Well Workers Interviews, Deckhand). Couples must develop ways of coping with the pressures, and talking about the problems contributes an important part of the answer. But sometimes it doesn't work.

CHAPTER FIVE

Job Satisfaction

What do fishers think about their jobs and how do they describe their working conditions?[1] Why do these men fish and what is it about this job that keeps them going back? Many fishers find it difficult to answer these questions:

Today is the 21st, my future is up to the 22nd. You live day to day. Anybody likes laying around for so long, but after a couple of weeks it gets pretty monotonous. I gotta get back at sea. It gets in your blood the same as a poker game, I guess. Same as anything else, it just gets in your blood. But one has to put up with it I guess. (Leavers Interviews, Deckhand)

In other studies deep sea fishers have described the three best features of their jobs as challenge and adventure, fellowship with co-workers, and earnings (Poggie 1980) – and the three worst features as health and safety hazards (Horbulewicz 1972; Poggie 1980), high stress (Horbulewicz 1972) and extended separation from home (ibid.). Job satisfaction relates to workers' perception of the value of rewards – both monetary and nonmonetary – for their work. Monetary benefits (e.g., wages, bonuses) may include compensation for the lack or loss of what fishers understand as nonmonetary benefits (e.g., adventure, independence). An inherent trade-off exists between these two types of benefits.[2]

A trade-off also exists between what benefits workers and what benefits employers. For example, as nonmonetary benefits decrease, employees may demand increased monetary benefits. Disgruntled workers may leave to work elsewhere in the industry or may leave the industry altogether. Even though employers may not perceive non-monetary benefits as obviously profitable, they must take such benefits into account. No labour process can function without the willing contribution of workers, and, in practical terms, companies provide nonmonetary benefits because management understands that people will not readily cooperate without them. Management must incorporate components of job satisfaction that reflect nonmonetary needs. They select work strategies that enhance positive aspects of the job, and, where possible, alleviate those aspects that impose the highest nonmonetary costs on employees.

Nonmonetary benefits appear to be critical components of fishers' job satisfaction. Anderson (1980) and Smith (1981) have cited fishing as an occupation where workers recognize job satisfaction as more important than monetary gain. Previous research (Anderson 1980; Smith 1981; Thiessen and Davis 1988) indicates that fishing has traditionally been an occupation pursued for its nonmonetary rewards – attachment to fellow workers and community, a sense of personal control, and the opportunity to work on the water. The erosion of these nonmonetary rewards leads to lower job satisfaction, especially when monetary rewards do not seem to compensate for their loss (Binkley 1990b; Thiessen and Davis 1988).

STUDIES OF JOB SATISFACTION
USING SURVEY DATA

Poggie and Pollnac (1978; Pollnac and Poggie 1979, 1988b) conducted the first studies based on survey data of fishers' job satisfaction. They looked at the relative job satisfaction of fishers based in various New England fishing ports, and they examined harbour-to-harbour variations in levels of job satisfaction. Their first study (Poggie and Pollnac 1978) included captains of seventy-nine small vessels working short trips out of Port Judith, Rhode Island, and forty-two captains of large vessels working long trips out of New Bedford, Massachusetts. Their second study (1979) included eighty captains from Maine working on small vessels.[3] Apostle, Kasdan, and Hanson (1985) replicated these studies for communities in south-west Nova Scotia. They surveyed 595 licensed fishers, representing 302 coastal, 126 deep sea and 167 coastal/deep sea workers (1985, 258).

In all of these studies, nonmonetary concerns, as well as earnings, represented important components in determining job satisfaction. Independence, adventure, and the challenge of the job were depicted as benefits. Negative factors directly related to working conditions, such as fatigue and mental pressures, living conditions on board, and long periods away from home, detracted from job satisfaction.

Except for the distinction between the coastal and the deep sea fisheries, researchers have treated fishing as an undifferentiated activity.

None of the studies mentioned above focused on variations in attitudes of workers with different job status or of workers on different types of vessels. Gatewood and McCay (1990, 1986) surveyed 391 commercial fishers from New Jersey who worked on vessels from fifteen to over a hundred feet in length (1986, 1, 7), surveying both captains and crew. The authors defined fishers as clammers, oystermen, scallopers, baymen (coastal fishers), longliners, and draggers. They used the same battery of questions as Apostle, Kasdan, and Hanson (1985), with seven additional attitudinal items. The analysis of their data focused on the differences between types of fishers (e.g., scallopers, clammers) and their job status.[4] Their general findings concur with the earlier studies mentioned above, but their research shows that job status and work organization influence the nonmonetary components of job satisfaction. They show that for the nonmonetary components, the higher the job status (e.g., captain), the greater the job satisfaction, and the more complex the work organization, the lower the job satisfaction.

METHODS OF ANALYSIS OF SURVEY MATERIAL

My 1986 and 1987 surveys included job satisfaction questions that replicated the twenty-two items used by Poggie and Pollnac (1978, 233–5; Pollnac and Poggie 1979, 3–6), and the twenty-six items employed by Apostle, Kasdan, and Hanson (1985, 257). (See chapter 1 for more details on the samples.) The responses to twenty-six items measured specific features of job satisfaction. These responses ranged from 1 to 5, with the greatest level of dissatisfaction being 1 and the greatest level of satisfaction being 5. I asked two questions to indicate general job satisfaction: "If you had your life to live over would you go into fishing?" and "If you had a son would you want him to be a fisherman?"[5] I also examined the relationship between those two questions and the twenty-six attitudinal questions.

Because of the large number of job satisfaction items, previous researchers have used two approaches to reduce the complexity of the findings. With the use of factor analysis (Apostle, Kasdan, and Hanson 1985; Binkley 1990b; Poggie and Pollnac 1978; Pollnac and Poggie 1979), the results do not appear entirely consistent, but they do suggest several stable dimensions such as Control, Earnings, and Work Quality.[6] Gatewood and McCay (1988) and Binkley (1990b) used another approach, imposing a theoretically informed classification to organize their job satisfaction items according to Maslow's (1954) hierarchy of needs. In this section we will employ both approaches in order to summarize patterns that emerge.

Factor analysis, a common statistical method, strives to classify a number of interrelated variables into a smaller number of dimensions or factors. In our case I used it to identify a small number of underlying factors contained in a larger set of questions about job satisfaction. I ran a factor analysis of the twenty-six job satisfaction questions to identify any underlying factors and to determine the relative strength of each question to each identified factor. First, I calculated the bivariate correlations (Pearson's r) between all items and then put them in matrix format. Then I fed this correlation matrix into the factor analysis procedure. Factor association depends on the common variations between the set of items. I assumed that those items that correlated highly with a factor could be represented by that factor. I interpreted the correlation between a factor and an item, referred to as factor loading, which ranges from 0 to 1, in the same way as a correlation coefficient. I constructed indices for these factors including any item related to a factor at the .40 level.

I also constructed three additional indices – Survival/Security, Belonging/Esteem and Self-Actualization – based on the social psychological work of Maslow (1954). This analysis used the twenty-six job satisfaction items and the method previously devised by Gatewood and McCay (1986, 10) to assign items to indices. I used scale analysis to eliminate any weak association between items and their corresponding index, and to ascertain the reliability of these indices. I applied Crombach's alpha to test the internal consistency of the job satisfaction and psychological indices.[7]

In order to understand the relationship between the job satisfaction and psychological indices identified, and the ethnographic material presented, I used one-way analysis of variance (ANOVA) to measure the levels of variation among job satisfaction and psychological indices by working conditions. This analysis identified those indices influenced by variation in working conditions, with three major variables being identified – level of capitalization, vessel and gear type, and job status. "Capitalization" had two categories – craft and industrialized. "Vessel type" had three categories – midshore, trawlers, and scallopers. "Job status" had two categories – captain/officers and crew. (See chapter 3 for definitions of these terms.)

I used data on job satisfaction to contrast the perceptions of and attitudes to working conditions of two groups of deep sea fishers. The first group, whom I called "stayers," included fishers originally surveyed in 1986 and who remained involved in the deep sea fishery. The second group, from the 1987 survey, whom I called "leavers," comprised persons who had left the deep sea fishery sometime in the five years prior to that survey. Those people still lived in Nova Scotia but worked in other fisheries, in fish plants, or in other industries, or collected social assistance (i.e., workers' compensation, unemployment insurance, or welfare).

In comparing the job satisfaction of fishers still employed in the deep sea sector with those who had left, I had to consider several factors. The recollections of those who had left the fishery might be coloured by the features of their current situation, such as their present employment status or present working conditions.[8] If, for example, their level of autonomy seemed greater than when they had fished, they would express less satisfaction with this component when evaluating their deep sea experience. Therefore I had to exercise caution in making any causal inferences. If those who had left expressed greater dissatisfaction with certain aspects of their deep sea job than those still fishing in that sector, this could not be assumed to have been a reason or cause for leaving deep sea fishing. It might signify nothing more than a reconstruction of the past in comparison with the present.

To emphasize the problematic nature of data from recollection, I postulated that former fishers would evaluate their deep sea experience less positively than current workers. Such a proposition has the merit of being congruent with expectations from social psychological theories such as cognitive dissonance, balance, and self-perception theory (Bem 1972; Festinger 1957; Taylor 1970).

In order to compare these two groups of fishers, I had to describe the current life situations of former deep sea workers. All of the former captains/mates still fished, many as independent owner-operators of smaller vessels. Most of the former officers also still fished. Many did not have sufficient capital or access to licences to become independent owner-operators; they served as crew for those small enterprises. Crew members who had left the deep sea worked in a variety of situations: many worked in land-based occupations, such as tire plants; some worked in other fisheries; others collected workmen's compensation or unemployment insurance benefits.

LEVELS OF JOB SATISFACTION IN THE NOVA SCOTIA DEEP SEA FISHERY

Table 5.1 summarizes rank order and mean levels of job satisfaction for all fishers in the study. These results indicate similar findings to those of the ethnographic material. Earlier studies also found high levels of dissatisfaction with the performance of federal and provincial officials. This item yielded a mean score below the neutral value of three. Note the low levels of satisfaction for the "time away" items –

Table 5.1
Mean levels of items and general indicators of job satisfaction for Nova Scotia
deep sea fishers in rank order

Item	Mean[1]	Standard deviation	Number of cases
l) Performance of federal and provincial officials	2.38	1.38	333
j) Time for family activities and recreation	3.06	1.51	334
t) Time away from home	3.14	1.40	334
a) Physical fatigue	3.77	1.24	334
g) Regular income	3.79	1.30	327
c) Mental pressure	3.82	1.23	332
o) Your earnings	3.84	1.30	332
v) Peace of mind	3.98	1.12	332
m) Time it takes to get to the fishing grounds	3.99	1.09	333
n) Adventure	4.00	1.10	330
z) Trip length	4.01	1.12	333
u) Opportunity to be your own boss	4.07	1.10	319
x) Cleanliness	4.07	1.19	333
q) Ability to come and go as you please	4.07	1.17	318
h) Hours spent working	4.14	1.11	333
r) Job safety	4.15	.94	333
k) Doing deck work on vessel	4.19	.89	327
s) Living conditions on board	4.22	1.03	334
p) Being out on the water	4.22	1.06	333
d) Healthfulness	4.26	.97	334
f) Challenge	4.30	1.00	331
w) Feeling you are doing something worthwhile	4.31	.89	333
b) Fellow workers	4.38	.88	334
e) Crowding	4.48	.85	333
i) Community in which you live	4.52	.86	332
y) Working outdoors	4.61	.65	332

General indicators	Yes	No	N
a) If you could live your life over, would you go into fishing?	197	126	323
	61%	39%	100%
b) Would you advise your son to go fishing?	50	262	312
	16%	84%	100%

[1] Responses for the attitudinal questions ranged from 1 to 5, with 1 indicating "very unsatisfactory" and 5 indicating "very satisfactory."

"time for family activities and recreation" and "time away from home" – which both yielded means of 3.1. In all deep sea samples to date, these two items have been identified as areas of low job satisfaction. For all other items, the level of job satisfaction appeared substantially above the neutral value but with wide variation within many of the categories. Throughout the current study, the second to the last item, "community in which you live," consistently ranked first or second,

Table 5.2
Factor analysis of job satisfaction items (varimax rotation)

Time		Control		Stress		Earnings		Adventure	
Time away from home	.81	Crowding	.55	Mental pressure	.74	Regular income	.81	Adventure	.78
Time for family activities and recreation	.61	Opportunity to be your own boss	.55	Physical fatigue	.52	Your earnings	.75	Challenge	.52
Trip length	.60	Ability to come and go as you please	.48	Fellow workers	.48				
				Hours spent working	.41				
				Feeling you are doing something worthwhile	.47				
				Cleanliness	.47				
				Community in which you live	.40				
Eigenvalues	6.1		2.1		2.0		1.7		1.6
Percentages	61.0		13.0		11.0		8.0		7.0

indicating the extreme importance of community attachment for these workers.

Although most of the workers (61 percent) would go fishing again if they had their lives to live over, only 16 percent would advise their sons to do the same. As might be expected, those workers who would not go fishing again seemed less content with all items than those who would go fishing. Those workers who would advise their sons *not* to go fishing had lower mean levels for all items than those who would advise their sons to go fishing.

Factor analysis for the whole sample identified the following underlying factors: Time, Control, Stress, Earnings, and Adventure (see table 5.2).[9] In order to compare our results with previous studies, I cite values for varimax rotation. Other rotations (equimax, oblique) yielded similar results. I rotated only those factors with an eigenvalue of one or more. By convention, the .4 level or above shows an important association (loading) between the factor and the question. I removed any item that did not "load" on a factor from the analysis. Table 5.2 specifies the factors ordered from left to right based on their eigenvalues – the percentage of variation involved in the factor analysis.[10]

Table 5.3 summarizes the job satisfaction indices constructed from the results of the factor analysis, and the psychological indices (with their corresponding components) with their corresponding Crombach's alpha values. All indices suggest a good level of internal consistency and reliability.

Items about work itself varied considerably depending on the general working conditions and job status of workers. Tables 5.4 and 5.5 respectively summarize the mean level of job satisfaction for the job satisfaction and psychological indices by working conditions. With some notable exceptions, almost all cases yielded statistically significant results.

Comparison of levels of job satisfaction for different vessel and gear type indicated that the workers on trawlers reported the least satisfaction with their work. The two indices most closely associated with community and family ties, Time and Belonging/Esteem, yielded substantially lower values for trawlermen. The index most closely linked to independence, Control, also yielded a significantly lower value for trawlermen. Except in relationship to Time and Control, scallop workers reported less satisfaction than midshore workers.

When comparing job status categories, crew members indicated less satisfaction than captains and officers for all indices. The Time and Adventure indices yielded almost the same results for both captains/ officers and crew.

Table 5.3
Job satisfaction and psychological indices

	Component items	Crombach's alpha value
JOB SATISFACTION		
Time	Time away from home, Time for family, Activities and recreation, Trip length	.79
Control	Ability to come and go as you please, Opportunity to be your own boss, Cleanliness, Crowding	.67
Stress	Physical fatigue, Fellow workers, Mental pressure, Doing deck work, Hours spent working	.75
Earnings	Your earnings, Regular income	.79
Adventure	Challenge, Adventure	.64
PSYCHOLOGICAL INDICES		
Survival/Security	Job safety, Cleanliness, Healthfulness, Mental pressure, Peace of mind, Living conditions on board, Performance of officials, Regular income, Your earnings	.73
Belonging/Esteem	Community in which you live, Fellow workers, Trip length, Time it takes to get to the fishing grounds, Opportunity to be your own boss, Ability to come and go as you please, Time away from home, Time for family activities and recreation	.79
Self-actualization	Working outdoors, Feeling you are doing something worthwhile, Doing deck work, Being out on the water, Challenge, Adventure	.73

Note: When using Crombach's alpha to test for internal consistency and reliability, a value of .60 is acceptable and a value of .70 is good (Marsh 1977, 254–5; cf. Apostle, Kasdan, and Hanson 1985, 284).

Table 5.4
Variation among job satisfaction indices by working conditions

	Vessel and gear type[2]				Enterprise size[3]			Job designation[4]		
	65 ft & under	Scalloper over 65 ft	Trawler over 65 ft	p^1	Large	Small	p^1	Captain/ officer	Crew	p^1
Time	3.7	4.1	2.9	***	3.6	3.6	NS	3.7	3.5	*
Control	3.6	4.0	2.6	***	3.4	3.5	NS	3.6	3.3	**
Stress	4.7	4.2	4.1	**	4.2	4.3	NS	4.4	4.1	***
Earnings	4.3	3.8	3.7	**	3.8	4.0	NS	4.0	3.6	**
Adventure	4.6	4.2	3.9	***	4.1	4.5	***	4.3	4.1	*

[1] As measured by analysis of variance (*F*-test) with probability levels indicated in the following manner: $*p \leq .05$, $**p \leq .01$, $***p \leq .001$.
[2] Number of missing cases was 3.
[3] Number of missing cases was 2.
[4] There were no missing cases.

Table 5.5
Variation among psychological indices by overall job satisfaction measures
and working conditions

	Vessel and gear type[2]				Enterprise size[3]		Job designation[4]		
	65 ft & under	Scalloper over 65 ft	Trawler over 65 ft	p[1]	Large	Small p[1]	Captain/ officer	Crew p[1]	
Survival/Security	4.22	3.76	3.73	***	3.78	4.01 **	3.94	3.75 **	
Belonging/Esteem	4.04	4.22	3.41	***	3.89	3.95 NS	4.04	3.78 **	
Self-actualization	4.61	4.31	4.06	***	4.21	4.50 ***	4.33	4.23 NS	

[1] As measured by analysis of variance (F-test) with probability levels indicated in the following manner: $*p \leq .05$, $**p \leq .01$, $***p \leq .001$.
[2] Number of missing cases was 3.
[3] Number of missing cases was 2.
[4] There were no missing cases.

When comparing levels of capitalization, those indices associated with the least tangible aspects of the work – Adventure, Self-Actualization and Survival/Security – appeared statistically significant. The other indices did not. With the notable exception of the Time index, satisfaction levels appeared higher for workers in craft enterprises.

I can summarize the findings of this study in the following way. Both the ethnographic and survey data show that fishers appeared

1 satisfied with their earnings, fellow workers, community attachment, adventure, and the challenge of their job;
2 less satisfied with items associated with increased outside control of their workplace; and
3 dissatisfied with the long hours away from home.

The survey data identified more specific variations; it indicated that

1 the more industrialized the fishery, the greater the dissatisfaction of workers with their working conditions and their earnings;
2 trawler workers reported the greatest dissatisfaction with all aspects of their work;
3 in comparison with midshore workers, scallopers reported greater dissatisfaction with their working conditions, except for time away from home;
4 midshore workers reported the greatest satisfaction with earnings;
5 crew members reported greater dissatisfaction with their working conditions than captains and officers; and
6 all workers reported very great dissatisfaction with the performance of government officials.

Comparisons with Other Studies
of Job Satisfaction

These results can be compared with the findings from other studies. Poggie and Pollnac (1978, 235–9) identified four factors for the Port Judith sample – Time, Outdoors, Earnings, and Independence – and identified the same four factors but added two more – Deep Sea and Work Quality – for the New Bedford sample. The combined sample of Rhode Island, Massachusetts, and Maine fishers identified three factors only – Time, Earnings, and Adventure (Pollnac and Poggie 1979, 4–6). (See table 5.6.) Apostle, Kasdan, and Hanson (1985, 259) identified eight factors: Independence, Work Quality, Earnings, Adventure, and Deep Sea; the New Bedford factor Time divided into two, Time/Family and Time/Trip length; and Crowding (a new factor).[11] Gatewood and McCay (1986) identified four underlying factors for job satisfaction similar to the Port Judith sample of Pollnac and Poggie (1979). In previous studies the factor here identified as "Living Conditions" was called "Offshore," the factor identified as "Stress" was labelled "Work Quality," and the factor identified as "Control" was called "Independence." Since this study focused only on the deep sea, the term "Offshore" did not seem appropriate.

Findings from the five studies yielded similar results, with minor discrepancies associated with the naming of factors and fluctuation due to sampling error. Variation in factor makeup stemmed from the interchange of "poor indicator" items that appeared to be associated with different factors. However, the basic pattern remained the same. Deep sea workers, in general, reported more dissatisfaction with the living conditions, quality of work on the vessel, the long working hours, and the time away from family and community than did coastal workers. Job dissatisfaction could be traced to the increased pace of the work schedule, the loss of independence, the lengthening of time at sea, and the erosion of community solidarity – all characteristics of the present industrialized deep sea fishery.[12]

The crews' growing dissatisfaction with deep sea work also makes it difficult for companies to get experienced and well-trained workers. In order to recruit the type of workers they want, they must address quality-of-life issues, and start offering both monetary and non-monetary incentives to obtain and retain experienced officers and crews. The deep sea fishery need no longer be a fishery of last resort.

The "unintended" consequences of these changes in resource allocation and work organization need not have been surprising. Earlier studies, from the late 1950s through to the early 1970s, had examined the organization of work on deep sea vessels and identified many of

Table 5.6
Questions defining the job satisfaction dimensions in the Nova Scotia and New England samples

Factor	Offshore Nova Scotia	Southwest Nova Scotia	All New England fishermen[1]	Point Judith, Rhode Island	New Bedford, Massachusetts
Control[2]	Opportunity to be your own boss .55	Ability to come and go as you please .72		Ability to come and go as you please .68	Opportunity to be your own boss .84
	Crowding .55	Opportunity to be your own boss .66		Performance of federal and state officials .68	Challenge .58
	Ability to come and go as you please .48	Being out on the water .45		Opportunity to be your own boss .65	Community in which you live .42
	Feeling you are doing something worthwhile .47	Feeling you are doing something worthwhile .40			
	Cleanliness .47				
	Community in which you live .40				
Work quality	Mental pressure .74	Mental pressure .66			Physical fatigue .71
	Physical fatigue .52	Physical fatigue .52			Mental pressure .63
	Fellow workers .48	Healthfulness .40			Performance of federal and state officials .63
	Doing deck work on vessel .44				Regular income .52
	Community in which you live .41				Community in which you live .44

Table 5.6
(continued)

Factor	Offshore Nova Scotia	Southwest Nova Scotia	All New England fishermen[1]	Point Judith, Rhode Island	New Bedford, Massachusetts
Earnings	Regular income .81 Your earnings .75	Regular income .90 Your earnings .82	Cleanliness .59 Physical fatigue .56 Regular income .49 Mental pressure .48 Job safety .45	Your earnings .68 Fellow workers .64 Regular income .54 Mental pressure .49 Physical fatigue .45	Your earnings .77 Ability to come and go as you please .70 Job safety .60 Fellow workers .57
Time/Trip length	Time away from home .81 Time for family activities and recreation .61 Trip length .60	Trip length .55 Time away from home .40	Time away from home .81 Hours spent working .72 Time for family activities and recreation .71 Ability to come and go as you please .61 Time it takes to get to the fishing grounds .47 Doing deck work on vessel .41	Time away from home .85 Hours spent working .61 Peace of mind .53 Time it takes to get to the fishing grounds .50 Crowding .50 Time for family activities and recreation .49 Trip length .45	Time away from home .83 Trip length .78 Hours spent working .74 Time for family activities and recreation .68 Time it takes to get to the fishing grounds .63 Doing deck work on vessel .54 Feeling you are doing something worthwhile .45

Adventure	Adventure .78 Challenge .52	Adventure .73 Challenge .54	Being out on the water .71 Adventure .71 Challenge .66 Working outdoors .57 Feeling you are doing something worthwhile .51	Working outdoors .80 Being out on the water .77 Adventure .55 Feeling you are doing something worthwhile .48 Healthfulness .42 Challenge .45	Being out on the water .77 Feeling you are doing something worthwhile .63 Adventure .59 Healthfulness .42 Peace of mind .41
Living conditions[3]	Living conditions on board .83	Living conditions on board .59 Cleanliness .58			Crowding .93 Cleanliness .83 Living conditions on board .63
Time/Family		Time for family activities and recreation .46 Time away from home .45 Hours spent working .45			
Crowding	Crowding .46				

Note: The numbers accompanying the questions are "factor loadings," or measures of the extent to which questions are involved in the particular factor.

[1] "All New England fishermen" includes the 80 Maine interviews. No separate factor analysis of the Maine data has been presented by Pollnac and Poggie (1979).

[2] Identified as "Independence" in the other studies.

[3] Identified as "Offshore" in the other studies.

these problems: health and safety hazards, extended length of time at sea, and high stress levels on the job.[13] Norr and Norr (1978) argued that variations in work organization within the modern fishery depended primarily on increased levels of capitalization that led to control of fishing gear by nonfishers. They suggested that where fishers retained control of the means of production, work organization would have pragmatic administration and nonbureaucratic structures. Similarly, where nonfishers controlled the means of production there would be a separation of major decision making from the direct influence of fishing conditions and crew interests, and a more bureaucratic structure.[14] They went on to hypothesize that these changes would lead to increased alienation of fishers in the industrial fishery, lower levels of job satisfaction, and a decreased labour force. Other studies (Apostle, Kasdan, and Hanson 1985; Gatewood and McCay 1990; Poggie and Pollnac 1978; Pollnac and Poggie 1988b, 1979; Thiessen and Davis 1988), based on analyses of the job satisfaction of fishers from Nova Scotia and the eastern seaboard of the United States, corroborated Norr and Norr's findings.

As noted above, job satisfaction can be measured in terms of monetary and nonmonetary benefits. Employers must be aware of the present levels of job satisfaction within the industry in order to weigh the costs of preserving nonmonetary benefits against the potential economic efficiency of new technology and other new management strategies. In order to maintain a source of labour in the future there must be a reassessment that includes the balance between monetary and nonmonetary benefits for workers and the economic benefits of technology and management strategies to the enterprise. The balance between monetary and nonmonetary benefits appears to be acceptable to most workers now involved in the deep sea fishery, but few would recommend this work to their sons. Moreover, companies find it harder to recruit and to retain young men in the deep sea fishery.

Comparison of Job Satisfaction of
Leavers and Stayers

Table 5.7 compares the mean levels of job satisfaction between current and former deep sea fishers. As expected, currently employed deep sea fishers expressed more satisfaction with many aspects of their work than did former deep sea fishers.

Distinctly different working conditions characterize the different occupational positions of captains/mates, other officers, and crew members. These social differences appear distinct enough to warrant

Table 5.7
Mean levels of job satisfaction among current and former deep sea fishers
by occupational position

	Captains/mates		Other officers		Crewmen	
Job satisfaction component	Stayed	Left	Stayed	Left	Stayed	Left
Physical fatigue	3.92*	3.00	3.82	3.29	3.67***	3.00
Fellow workers	4.46	4.45	4.34	4.64	4.34	4.46
Mental pressures	3.70*	2.82	3.88	3.43	3.87***	3.07
Healthfulness	4.41	4.09	4.34	4.29	4.14	3.80
Crowding	4.59	4.27	4.47	4.46	4.43***	3.77
Challenge	4.57	4.55	4.13	3.93	4.21	3.81
Regular income	4.07*	3.27	3.88	3.73	3.59	3.33
Hours spent working	4.20***	2.91	4.24	4.00	4.07***	3.29
Community in which you live	4.56	4.71	4.52	4.40	4.49	4.54
Time for family activities and recreation	3.51	3.13	3.03	2.58	2.82	2.71
Doing deck work on vessel	4.24*	3.55	3.83	3.69	4.29	4.09
Performance of federal and provincial officials	2.40	2.40	2.58*	3.07	2.40	2.82
Time it takes to get to the fishing grounds	3.91	3.27	4.01	3.75	4.02**	3.52
Adventure	4.37	3.91	3.45***	4.42	4.00*	3.63
Your earnings	4.20	3.82	3.84	3.67	3.63	3.64
Being out on the water	4.40	4.36	4.06	4.40	4.18	4.30
Ability to come and go as you please	4.38**	3.36	4.34	4.00	3.80	3.47
Job safety	4.39*	3.00	4.12*	3.13	4.01***	2.55
Living conditions on board	4.46	4.00	4.16**	4.71	4.09	3.88
Time away from home	3.39**	2.09	3.25	2.50	2.95	2.68
Opportunity to be your own boss	4.52	3.55	4.23*	3.50	3.71***	3.02
Peace of mind	4.20*	3.30	3.85*	2.93	3.90***	3.20
Feeling you are doing something worthwhile	4.51	4.09	4.34	4.27	4.18***	3.56
Cleanliness	4.35	4.09	4.09***	4.79	3.89	3.68
Working outdoors	4.65	4.82	4.38	4.33	4.67	4.63
Trip length	4.21*	3.50	4.07	3.57	3.86*	3.45
Minimum N	99	11	67	14	156	51

*$p \leq .05$; **$p \leq .01$; ***$p \leq .001$.

separate explorations of patterns of job satisfaction. Moreover, previous research has shown that occupational status influences the working conditions and consequent job satisfaction (Gatewood and McCay 1988). Therefore, I will examine each of the occupational positions separately. The first two columns of table 5.7 provide the mean level of satisfaction reported by current and former captains/mates. The patterns here appear straightforward: first, all statistically significant differences (which appear on ten components) conform to the expected pattern – former captains and mates expressed less satisfaction than current ones. Former captains/mates described deep sea fishing as particularly stressful, with significant differences on "job safety," "physical fatigue," "mental pressures," "peace of mind," and "trip length." They also recalled having limited autonomy, scoring lower on "ability to come and go as you please." Finally, they appeared less satisfied with "doing deck work on vessel" and "regular income."

Researchers considered absence from friends and family to be the most important drawback in the fishery. Our findings support this view: "time away from home" and "hours spent working" received the lowest ratings among captain/mate leavers, with mean scores less than the neutral point of 3.0. In addition, former deep sea captains and mates reported significantly lower satisfaction with both "time away from home" and "hours spent working" than did current ones. Given the findings discussed in the previous chapter on families, it is apparent that for many fishers the dissatisfaction of being "away from home" did not mean dissatisfaction with being away from their wives and children, since many fishers spent much of their time on shore in the company of other men. However, when these fishers' wives said they were dissatisfied with their husbands being "away from home" they meant they were dissatisfied with their husbands being away from them and their children. This mismatched expectation is the source of many family conflicts.

Those job satisfaction components that do not differentiate between stayers and leavers provide additional insights. First, both groups reported strong community and crew attachment. The job satisfaction scores for "fellow workers" and "community in which you live" remained as high for the leavers as for the stayers. Second, both current and former deep sea fishers reported great satisfaction with the environmental aspect; "working outdoors" and "being out on the water." Third, levels of satisfaction with both the general "living conditions on board" and the specific "cleanliness" seemed about the same for the stayers and leavers.

In contrast to the captains and mates, levels of job satisfaction of current and former "other officers" appear more complex. For this occupational position, workers reported seven statistically significant differences, but in four of these cases former officers reported greater satisfaction than current ones. Current officers reported greater satisfaction with "job safety," "peace of mind," and "opportunity to be your own boss." Former officers recalled greater satisfaction with the general "living conditions on board," and specifically with "cleanliness."

They seemed to miss the "adventure" of their former jobs, recalling greater satisfaction with this than do current officers. Finally, on the issue of "the performance of federal and provincial officials," the leavers recalled less dissatisfaction than the stayers.

Fishers have universally targeted the performance of federal and provincial officials as the object of their greatest dissatisfaction (Apostle, Kasdan, and Hanson 1985; Binkley 1990b; Thiessen and Davis 1988). No researcher has reported variables or subgroups of variables positively associated with fishers' attitudes to the performance of federal and provincial officials.[15]

The final two columns of table 5.7 contrast current and former crew members. On twelve components, former crew members recalled significantly less job satisfaction than did current crew members. As with the captains, former crew members recalled no items with significantly greater satisfaction than did current crew members. Satisfaction with the physical and mental demands of deep sea fishing (i.e., "physical fatigue," "crowding," "hours spent working," "job safety," "trip length," "time it takes to get to the fishing grounds," "mental pressures," and "peace of mind") clearly differentiated stayers from leavers. In addition, current crew members valued more highly the "challenge," sense of "adventure," and "feeling you are doing something worthwhile" than did former crew.

A close examination of Table 5.7 suggests that some common components of the work environment differentiated stayers from leavers irrespective of their occupational position: stress – both physical and mental; the time demands of the job; health and safety hazards of deep sea fishing; and the limited independence in the deep sea sector. To explore this more systematically, I constructed indices to measure these aspects of job satisfaction (see Table 5.8).

On the first four factors listed in Table 5.8, leavers reported significantly less satisfaction than stayers. This applied not only for the sample as a whole, but for each of the three categories of occupational status. This consistency suggests a common dynamic that links features of the work environment to ceasing to do that type of work. I suggest that the fishing schedule – long trips with forced absences from social familiars – combined with the limited control deep sea fishers have over their work schedules creates high stress levels. Under these conditions accidents occur more often, forcing some fishers out of the occupation. Not surprisingly, in retrospect these leavers expressed greater concern about health and safety matters.

The fifth factor, "earnings," does not indicate statistically significant differences between stayers and leavers. Former fishers recognised that in economic terms, deep sea fishing compared well with the few

Table 5.8
Mean levels of job satisfaction among current and former deep sea fishers

Index	Job satisfaction component	Current workers	Former workers
Control	Ability to come and go as you please, Opportunity to be your own boss	4.07*	3.40
Time	Trip length, Time away from home, Time it takes to get to the fishing grounds, Time for family activities and recreation, Hours spent working	3.67*	3.11
Safety/Health	Healthfulness, Job safety	4.21*	3.34
Stress	Peace of mind, Physical fatigue, Mental pressures	3.85*	3.07
Earnings	Your earnings, Regular income	3.82	3.57
Adventure	Challenge, Adventure	4.15**	3.86

$*p \leq .01; **p \leq .001.$

available alternatives. Finally, leavers, particularly crew, tended to report less satisfaction with the "adventure" component of deep sea fishing.

In our description of the two samples, I mentioned that former fishers were, on the average, older than those still fishing. This raises the possibility that differences in levels of satisfaction reflect a function of age rather than turnover status. To test this possibility, I computed Pearson's correlations between age and each of the six factors. Age has a statistically significant relationship regarding satisfaction with "earnings" and "safety." In both instances, older fishers reported greater satisfaction. Since satisfaction with "earnings" and turnover status appeared unrelated, this first finding seems irrelevant to the issues raised here. Furthermore, older fishers, who were overrepresented among the leavers, expressed greater satisfaction with safety. In terms of our survey, the effect of age suppressed the strength of the relationship between turnover status and satisfaction with safety. In other words, the difference between stayers and leavers concerning satisfaction with job safety would likely be somewhat stronger than reported if the two groups had had similar age distributions.

Similarities and Differences in Job Satisfaction Profiles

I also used a third technique, Q-correlations, to assess the similarities and differences in the job satisfaction profiles. Using this approach,

Table 5.9
Items showing the least job satisfaction by turnover status and occupational position

Position	Current workers	Former workers
Captains and mates	Performance of federal and provincial officials	Time away from home
	Time away from home	Performance of federal and provincial officials
	Time for family activities and recreation	Mental pressures
	Mental pressures	Hours spent working
	Time it takes to get to the fishing grounds	Job safety
Other officers	Performance of federal and provincial officials	Time away from home
	Time for family activities and recreation	Time for family activities and recreation
	Time away from home	Peace of mind
	Adventure	Performance of federal and provincial officials
	Physical fatigue	Job safety
Crewmen	Performance of federal and provincial officials	Job safety
	Time for family activities and recreation	Time away from home
	Time away from home	Time for family activities and recreation
	Regular income	Performance of federal and provincial officials
	Your earnings	Physical fatigue

the "cases" represent the twenty-six job satisfaction items and the "variables" denote the combinations of occupational position and turnover status.[16] This technique permits us to assess the similarity/ difference of job satisfaction profiles of various subgroups of stayers and leavers.

When developing these profiles I first computed the five "worst" and the five "best" aspects of deep sea work as reported by stayers and leavers in each occupational position – see tables 5.9 and 5.10 respectively.

Among current workers (stayers), the "performance of federal and provincial officials," "time away from home," and "time for family activities and recreation" denote the three worst aspects in each of the occupational positions (see table 5.9). Two of these – "time away from home" and "performance of federal and provincial officials" – also appear among the five worst components for all three groups of

Table 5.10
Items showing the most job satisfaction by turnover status and occupational position

Position	Current workers	Former workers
Captains and mates	Working outdoors	Working outdoors
	Crowding	Community in which you live
	Challenge	Challenge
	Community in which you live	Fellow workers
	Peace of mind	Being out on the water
Other officers	Community in which you live	Fellow workers
	Crowding	Crowding
	Working outdoors	Adventure
	Fellow workers	Community in which you live
	Healthfulness	Working outdoors
Crewmen	Working outdoors	Working outdoors
	Community in which you live	Community in which you live
	Crowding	Fellow workers
	Fellow workers	Being out on the water
	Doing deck work on vessel	Doing deck work on vessel

leavers. As one might expect, "job safety" appears among the worst five aspects for all three groups of leavers, but not among the worst five aspects for any of the three groups of stayers.

Turning to those items that provide the greatest satisfaction, "working outdoors" and "community in which you live" appear among the top five aspects in all six groups (see table 5.10). In addition, "crowding" appears among the five best components for all groups of stayers, whereas "fellow workers" appears among the five best components for all groups of leavers.

In many respects the working conditions of deep sea workers seem similar to those of the fishers on scallopers and draggers, and to some extent to those of the workers on longliners studied by Gatewood and McCay (1988). The three components of job satisfaction rated lowest by our current workers ("performance of federal and provincial officials," "time away from home," and "time for family activities and recreation") appeared among the six worst components for those fisheries reported by Gatewood and McCay (1988). Likewise, "working outdoors" and "community in which you live" appear among the most satisfying items for both our stayers and Gatewood and McCay's scalloper and dragger workers (ibid.). This also corroborates a tendency documented by Apostle and co-workers (1985), that attachment to community and to fellow workers appears particularly pronounced among Nova Scotian fishers.

Table 5.11

Job satisfaction profiles combined with turnover status and occupational position (Pearson's Q-correlations)

	Current workers			Former workers		
Current workers	(1)	(2)	(3)	(4)	(5)	(6)
(1) Captains/mates	—					
(2) Other officers	.88	—				
(3) Crewmen	.88	.88	—			
Former workers						
(4) Captains/mates	.77	.66	.77	—		
(5) Other officers	.66	.66	.72	.80	—	
(6) Crewmen	.61	.58	.75	.87	.80	—

In only one instance did an item – "adventure" – appear among the worst five aspects for one group and among the best five aspects for any other group. Former "other officers" reported "adventure" among the top five aspects of their work and current "other officers" reported it among the bottom five.

The findings discussed so far suggest that both occupational position and turnover status influence job satisfaction profiles. To explore this more systematically, I created six groups by combining turnover status with occupational position (captains/mates, other officers and crew) and used Q-correlations. In such an analysis, the twenty-six components of job satisfaction represent the "cases," while the mean job satisfaction scores for the six groups denote the "variables." I computed the Pearson's correlations on a twenty-six by six matrix where the cell entries incorporate the mean job satisfaction scores on a given component (row) for a given group (column). Such a procedure ignores absolute differences in levels of job satisfaction among the groups but captures relative differences. It answers the question, "To what extent are those job satisfaction components that one group rates highly also rated highly by another group; and, conversely, to what extent are those that one group rates low also rated low by another group?"

Table 5.11 shows the job satisfaction similarity profiles in the form of Pearson's correlation coefficients among the six groups.[17] Three patterns emerge:

1 The profiles of the three groups of current deep sea workers most closely resemble each other, with all three possible correlations identical at .88;

2 Likewise, the profiles of the three groups of former deep sea workers most closely resemble each other, with correlations ranging from
  .80 to .88;
3 In every instance, the profiles of groups with like turnover status but
  unlike occupational position more closely resemble each other than
  any groups of like occupational position but unlike turnover status.

TIPPING THE BALANCE

The differences in the profiles and patterns of levels of job satisfaction
of current and former workers indicate that the combination of long
working hours, the long time away from home, the loss of autonomy
while working, the physical and mental stress of the work, and safety
concerns represent common areas of dissatisfaction for all former
fishers. How these concerns materialize depends partly on the occupational position of the fisher. The link of having left the deep sea
seems to produce more common concerns than the link of occupational position. The configurations of satisfaction/dissatisfaction
reported among former deep sea workers of different occupational
positions more closely resembled each other than the configurations
of stayers and leavers who held identical occupational positions.

Two features of deep sea fishing remain at the heart of the problems
examined in this chapter: the harsh working conditions and concomitant work schedule, and the loss of workers' control over the work
processes – both on the deck and in the wheel-house. Fishers, rugged
individualists, want to make their own decisions at sea and to rely on
their own skills and knowledge in making these decisions. But, increasingly, management or technology determines these decisions. The lack
of control over the technology used and the demanding work schedule
make for a dangerous combination that results in high stress and
increased accident rates.

CHAPTER SIX

Trading Off Wages and Nonmonetary Benefits

GENERAL WORKING CONDITIONS

Fishers understand the trade-offs they make between their wages and the nonmonetary benefits of their job. Earnings and making a good living remained very important to most fishers interviewed. Time and time again they emphasized that few jobs open to men with little education paid so well. (See chapter 1 for the income ranges of those surveyed.) They recognized that the deep sea fishery, with its long working hours in poor working conditions, exemplified a young man's job. Fishers discussed at length their actual working conditions on board the vessels. They accepted the risks involved in their work. Many compared the risks that they ran at sea to the risks when driving a car. Although most men saw their work as hard physical labour and accepted broken or extended watches as common, they also played down the long hours and concomitant fatigue. As a number of men said, "a man can work eighteen hours easily," and all proudly told stories of the longest shift they had ever worked. Watches could be fifty or sixty hours, but the long hours always meant a large catch and a big pay-cheque. For most though:

It's just a job. You just watch what you're doing and everything runs so smooth. Oh, some are slave drivers you know, but these are the guys that make good money. They have good crews. But the crews tend not to stay with them that long because the skipper tries gettin' more out of the man's body than what he's capable of. But now that we have our dunk tables and everything, it's running pretty smooth. But we still have our slave drivers. You don't get more accidents on boats like that, not normally. Just a lot more tired men, that's all. (Well Workers Interviews, Deckhand)

When we asked fishers what they liked about their work, they inevitably spoke about the adventure and challenge of the job. They liked the ability to come and go as they pleased. A fisher likes being his own boss.

Freedom – it's just freedom. You have no bill collectors down at your door or nobody's selling fifty cent gospel papers. If you're married, there's no nagging. There's not somebody down your back all the time. There's peace of mind. When you do come home you enjoy it a lot more. Too much bad weather, winter time, and a miserable skipper – that's the bad side, but other than that it's work. Everybody works. When you're young you're taught that fishing and luck is the wet ass and a hungry gut. But you take the bitter and the sweet. You try and fill the holds in your boat up but if you can't, you can't. (Well Workers Interviews, Deckhand)

Whom they work with appears extremely important to fishers. Many men spoke of leaving a vessel because they did not get along with other crew members. Most of the time, the crew function just like family, and fishers do spend more time with fellow crew members than with their own families. Because of the inherently dangerous work on board, crew members look out for each other. Here, a fisher describes how they take care of each other:

Having good crew is beautiful. A good crew works and sticks together. They can protect each other, watch out for each other, and everything runs so much smoother. Morale is kept up. The first thing you know you got your hold full of fish and you're headed back home. It's time to go and have a beer together and have a laugh until the next time. To have a good crew – it means a lot. After awhile you don't know when you step aboard. You work with them for one or two trips and you'll find out if they're okay or not. You always are bound to get one or two guys that you don't like. I guess they don't like you either. But you try to cope with these people.

There's the odd guy that's kind of careless, hyper, stuff like that. So if he makes a little mistake he don't notice, you go behind and correct him. But you don't want to do it in front of the man's face. It's very important that everybody digs in like and does what's right. (Well Workers Interviews, Deckhand)

The attitude of the captain represents a crucial factor in a worker's earnings, job satisfaction, and safety. One fisher summarized his feelings about skippers this way:

The skipper – he's number one man. If he is a good skipper, he has his bad days and good days too. But if he is a good man, he can handle his crew by just being nice to 'em. You know, I'll be nice to you, you be nice to me. This is the way it works. But if you get a skipper that's hollering and cursing you, calling you an old woman and stuff like this well, well he's lucky to see the next stop. So we don't have any troubles. The last ones that I've been sailing to, they're good men. A good captain draws a good crew and a good captain will take up for his crew. He'll make sure that the good crew makes a fair living. (Well Workers Interviews, Deckhand)

EFFECTS OF MANAGEMENT CONTROL

In general, management practices associated with the industrialized fishery and the enterprise allocation have led to a more bureaucratic workplace. No matter how sympathetic the skipper or how good the crew, the organization of the work process has eroded workers' sense of personal control, dehumanized the workplace, and isolated crew members from the decision-making process. Crew feel alienated from the officers and captain of the vessel, as well as from the management of the company. Here's how one fellow put it:

There's adventure in fishing if you go about it in the right way – if you go fishing on your own. Now them draggers have so many people and the company's got to pay your wages. You know that they have to have the most of everything you got. The adventure is for the guys who make the decisions. Well, you should be able to come and go as you please but you don't. You got to come when they tell you. You gotta go when they tell you. Now where's the fun in that? (Well Workers Interviews, Deckhand)

Fishers, particularly captains and officers, value a sense of independence, a challenge, and the feeling of being their own boss. As companies increase their ability to control operations from shore – through better communication between the vessels and head office – job satisfaction among the officers declines. Companies instruct captains according to the needs of plants and markets. Captains feel pressured to check with the company for approval of routine decisions, and often the captain sails without knowing where the vessel will land. They also feel that the land-based company official in charge has neither the experience nor the understanding of the situation to make an informed decision about what may be appropriate at sea. This perception frustrates the captain, who feels he has the experience to do the job without someone always looking over his shoulder. This loss of control appears critical to captains, who refer to those trips as "Sobey's trips" or "supermarket runs." (Sobey's is the name of a regional supermarket chain.) They talk about just "going shopping" and sense a lack of challenge in the job. It erodes their feeling of doing something worthwhile and makes them feel like a cog in the machinery.

Under this system, as in the past, earnings for workers depend on catch size and fish type, but fishers' incomes have stabilized and become more uniform because catch size and fish type are known ahead of time. Often fishers earn more or less the same income as they did under the previous quota system, but sometimes, for instance in the case of highliner crews,[1] their income has declined – in some cases dramatically. While this system decreases the economic risks to workers, it also decreases the possibility of economic benefits from good fortune and skilful fishing.

Crew members can increase their earnings only by catching particularly highly valued species or by getting bonuses for quality. This example will illustrate the point:

A lot of our fishermen got pushed too far. I had a mate with me for years. In fact, he was mate with me for quite a while and he got a chance to skipper a vessel. That man has done perfect. He went out all winter cod fishing. Then went out shrimping. He's gone twenty-seven, twenty-nine days and he got a load of shrimp. Christ, his crew made more money than I made in the last six months. I think they made eleven thousand dollars in twenty some days. I mean that's a special trip, shrimp. I'm not saying that. His company told him they'll be paying him, I think, forty some cents a pound for his codfish. He'll be foolish not to go. If he stays with our company he'd be getting sixteen, seventeen, and eighteen cents a pound. There he can get forty some cents a pound. (Well Workers Interviews, Captain)

Fishers have no control over the time at sea. The needs of the plants have become paramount, resulting in uncertainty about where off-shore crews will be fishing and where they will land. When companies have almost reached their allocations, there may be very little money to be made for the crew. Workers receive earnings based on a share system with bonuses for high-quality fish, and if the species of fish and the size of the catch remain constant, then they earn more for a shorter trip at a perceived higher rate of pay. Even when they catch the specific amount of fish needed by the company, a vessel may bide time at sea until the receiving plant can accept the shipment. Here's how one crew member described the effect of these policies on his vessel:

We was out there high as seventeen days [on the groundfish trawler]. We was suppose to land on our fifteenth day, that's when everything was top secret. They wouldn't tell the skipper nothing. If they was going to stay an extra day, they wait till the last minute. There you are on this watch and the next watch when I come up, we be cleaning up, we be going home. Then we'd come up and have to shoot away because we were going to stay an extra day. But now it's pretty good the way they got it. It's not good but it's a lot better than what it was. (Well Workers Interviews, Deckhand)

Time at sea greatly influences earnings in other ways. The longer the harvested fish remain at sea, the poorer their quality. In these cases, workers feel they have no control over their pay schedule and no way to increase their earnings. This attitude leads to frustration and bickering. Sometimes the worker leaves the deep sea for the apparently more lucrative midshore and coastal fisheries.

The drive for improved quality has resulted in the use of new and different technologies. These technological advances have made fishing more efficient and profitable, but pay little attention to human limitations. Most of the changes represent labour-intensive rather than energy-intensive procedures. For example, instead of relying on the pen system, companies have introduced containers to store the fish. Boxing, a labour-intensive method, requires lifting plastic boxes that contain either 70 or 110 pounds of fish and ice. Crews work in a refrigerated hold with conveyor belts overhead; some of the boxes have been filled with ice before leaving port, and the ice must be chipped and shovelled from one box to another as the catch fills the hold. The fish arrive via the conveyor belts from the fishroom – where they have been washed and cleaned – and land in the prepared boxes. Workers put ice on the bottom, in the middle, and on top of the fish. The full boxes must be lifted and carried while walking on the narrow ledges of other boxes, and stacked in rows. In a rough sea these unsecured boxes may come tumbling down on the workers. Even in good weather, the work floor heaves and rolls. The men stacking the boxes have no control over the rate at which other men fill the boxes. If someone falls in the fish hold, the whole process of gutting and preparing the fish and filling the boxes must also be stopped. A captain described the following problems with boxing:

It's hard to keep hold of the containers in the hold. People are lifting the boxes, the boat is rolling, and they wrench their back and what not ... One of the problems is that they have to stand on top of one row of the fish boxes to pile the next. Whenever you're standing on plastic with ice on it, it's got to be slippery. Then, in a real rough day, I guess the boxes move a certain amount and sometimes they roll over, you know. Somebody could have a bad accident ... The stacking and the way the people had to work, I really wasn't very impressed. Those little ridges ... have you ever seen the boxes? Those little ridges on top, about an inch wide – there is not very much to hold another box, is it? The whole stack shifts. I mean you get out in a bad breeze, and a rolling ship, and it is pretty hard to keep something from moving that's not really fastened down. It's pretty hard to keep it solid. I've heard tell of a couple of boats that had a double stack fall over. Fellows like us coming up from Labrador in wintertime four days steaming, you don't know what you're going to face before you get to Nova Scotia, do you? There wouldn't be nobody in the hold then, once you leave to come home that's it. What I mean is you could be working in the hold, and there could be stuff flying around when it's rough. You fish pretty bad weather, you've got to. You can get somebody hurt. (Well Workers Interviews, Captain)

Clearly a redesign of some of the technology according to ergonomics is essential. Some Scandinavian research focuses on these problems (H.G. Andersen, personal communication, May 1989).

The majority of workers feel that the harder work associated with these technologies leads to better-quality fish and profit for the company. They also believe that the company does not pass along those advantages as increased earnings to the workers who harvested the better-quality fish.

How long a vessel spends at sea depends on several factors: the number of hauls, the technology involved, the availability and maintenance of both fish-finding and fish-catching equipment, the frequency of equipment breakdowns, the experience of the captain, and the availability of fish, to name a few. Since workers want to get home as fast as possible and to make as much money as possible on a trip, they will push themselves when the fishing is good. They will work beyond their regular six-hour work shift, and push their equipment to the limit in order to bring in large catches of fish. But fatigue leads to higher accident rates.

If the catch declines, if the technology breaks down, or if the preponderance of the wrong type of fish results in low catches or a long time out at sea, then fishers expect low pay and become disheartened. That can happen at any time, but it occurs most commonly when the allocation for a species has almost been reached, or at the end of the season for a particular species in a particular area. Fishers say that the number of such trips has increased under the enterprise allocation system. They feel that the company should recognize those problems and reward the workers for taking that type of trip, since it consistently results in low pay for the crew.

In deciding which vessel and crew fishes where, and for what species of fish and how many of them, the company indirectly sets the income for the officers and crew on each vessel. In order to get the best income for their crews, captains now feel that they must compete with each other to get the most lucrative assignments for their vessel.

Instead of open competition between skippers for the most fish or the highest landed value, the captains compete for the ear of the company dispatch officers who set the trip schedules and routes.[2]

Fishers' most frequent complaint was too much time spent away from home, and therefore anything that increased time at home and decreased time at sea would be good for morale. Companies strive to apply a standard schedule of eight to ten days at sea alterating with forty-eight to seventy-two hours at home. In principle trip schedules should be more regular now, because managements can more precisely predict and anticipate the needs of their plants. However, sometimes the vessel must stay out until the shore plants need the fish. In those circumstances the workers become bored. They do fire drills, clean the vessel, play cards, or watch the VCR. Captains worry about the morale of their crews, and fear an increase of drug and alcohol use on board. With large companies, the frequent redirection of vessels to ports other than the home port, with a long journey home or else a sojourn in the receiving port, also erodes workers' job satisfaction.

Many fishers complain about the short turnaround time they have on shore:

The company sets the work schedule. The company that I've been working for, we have five days in port. But the bigger companies, they're sending their men out two and three days after they come in. They are tired. This could create more accidents than what is normal if you could have your rest ... They're just tiring the men out more. It's greed. (Well Workers Interviews, Captain)

That kind of comment about the drive for profits persists. Unions have been able to increase fishers' earnings, but at the cost of non-monetary benefits. Some fishers blame the unions, others blame the companies, but all agree that lengthier trips and shorter periods at home result in exhausted men.

Workers on those large vessels have little control over their schedules, and most spend their off time "recovering" from the work shift, particularly if they have to travel a substantial distance from their home community to the ship's home port. The effects of this stress can build up over time, and fishers speak of feeling like an animal let out of a cage, or of being broken up. A common response to this stress is to go on a "tear":

What I find is you're always trying to catch up – make up for lost time. Trying to live in forty-eight hours what the guy ashore does in ten days or two weeks. Drink, drive, run around, watch TV, there's no time for sleep. You have to keep going, going all the time. (Well Workers Interviews, Deckhand)

Such a demanding schedule creates both physical and mental stress, which the fisher tries to dissipate through onshore activities with other men. His spouse resents these activities – drinking, driving, running around, watching TV – and this resentment adds an additional layer of stress. Company limits on the number and kind of allowable deferments of trips make this situation worse, particularly for families. As one fisher's wife expresses it,

With the union they're given certain times home: Boxing Day, New Year's, they used to be gone. Before you had children you'd look forward to New Year's, but they didn't have it, they had nothing. Now they stay home Christmas and New Year's. But other family times, Mary's graduation for example, he had to lose a trip ... First communions, baptisms, all the family things you want to do together, he has to take a trip off ... The contract gives you three trips off. So you take one for graduation and another if someone is ill, come summer when we want to be gone, he's still there. And then it's winter. (Wives Interviews)

These workers have little time to become involved with their wives and children. They see the families they go to sea to support functioning without them. They wonder what part they play in their lives. As one man said:

How does fishing affect the family? Ah, if you're away from your family for two weeks and you're home for one night out of a month – I ask you, how would you feel about it, if you was a woman? It's hard on the children. It's hard on the mother. It's hard on the father. It has to be, seeing that the family has but one day a week. You're passing the pay-cheque and that's it. Or one day a month. You know it's outrageous. Your kids grow up and you don't actually know them – that's definitely. You can come home, and a guy will be going down the street with them, and the kid would say "Hi Daddy," and you wouldn't know the difference. You're just the stranger that comes home and buys them a little gift once and a while. (Well Workers Interviews, Deckhand)

The typical worker takes three trips off: one to go deer hunting, one or two to go coastal lobstering, and one for a family holiday in the summer when school has closed. Men have little time to develop relationships with their spouse, children, or other family members and friends, or to develop social and recreational interests. This lack of participation undermines community solidarity and alters the nature of community life. The resulting stress can surface as drug and alcohol abuse, family breakdown, and child or spouse abuse.[3]

To what extent is this work schedule intrinsically necessary to maintain an efficient deep sea fishing industry? Part of the answer lies in the vertical integration of deep sea fishing companies. Because of enterprise allocation, companies must maintain strict control of harvesting in order to coordinate it with processing needs. We can easily see the sources of the difficulties, but not the solutions. Interviews with the fishers made it clear that the problem was not the length of time at sea, but rather the short time on shore. However, they stated categorically that they would be unwilling to take a lower income in order to have longer shore leave, and the companies appear unwilling to offer the same income for less work.

The enterprise allocation and associated business strategies entered the deep sea fishery to make fish production more responsive to the market, rather than reactive to undifferentiated harvesting. Those changes produced a mixture of both anticipated and unanticipated benefits and unforeseen disadvantages. Benefits included the increased feasibility of scientific management, less uncertainty about the share of fish caught (allocation), the increased quality of fish, the increased safety related to stability, and the regularization of income, sea time, and work schedules.

In spite of the benefits to fishers, the needs of the men on board remain secondary, and the demands of the plants, which respond to market pressures, prevail. These features combined with an increasing lack of control of the workplace have led to workers' dissatisfaction and, in some cases, to the exodus of skilled workers from the fishery. The increase in accidents not only stems from the new technology introduced onto the vessels but also from those losses of skilled workers. Neither companies nor employee groups have fully appreciated or investigated these effects. Harvesting at sea represents a very different form of production from working on land and the differences produce problems that must be addressed. The solution appears complex and difficult.

SAFETY CONCERNS

In the deep sea fishing environment, part of management control includes safety regulations. A safe environment reflects a controlled work environment. A safe environment means fewer accidents, less down time, and reduced expense, making safety as much a question of economics as a question of workers' well-being. The companies' response to increased accident rates has been to establish greater control. A safe vessel has minimal hazards and has a captain in control of the ship, its environment, and its crew. But we must go beyond the issue of control and look at the underlying causes of accidents. The environment at sea can never be fully controlled; unpredictable weather and the mobile resource mean production will never be fully under the captain's control.

By implementation of the enterprise allocation and its associated business strategies, companies hoped to lessen the likelihood of accidents such as listing or capsizing due to overloaded vessels. Since management specifies the amount of fish to be harvested, and the "daily hails" allow the company to know when the specified catch has been or will be taken, the company can bring the vessel in to the appropriate plant as soon as possible. This procedure has virtually eliminated problems associated with critical stability.

The drive for quality purported to also reduce accidents associated with "speed-up" (increasing the rate of work by lengthening normal shifts and shortening rest periods to take advantage of good fishing) but Workers' Compensation Board statistics suggest an increase in accidents and injuries. The reason for breaking watches and speeding up processes had been competition for the resource at sea, and once this competition stopped, accidents and injuries should have decreased. However, since fish remain a mobile resource, a captain cannot be assured of finding schools of fish when he wants them. He continues to try to take as much of the resource in as little time as possible, which results in broken watches and speed-up.

Accidents attributable to boxing fish include injuries due to lifting (e.g., back problems, hernias), slips and falls (e.g., sprains, strains, and fractures of limbs), and being hit by falling boxes (e.g., crushing of body parts). Illnesses associated with the cold and damp working conditions have also increased (e.g., rheumatism, colds).[4]

Extreme variations in working conditions produce fatigue (from speed-up and broken watches) and boredom (on the long steams to and from the fishing grounds). When fish stocks decline, frustration over lost wages leads to further worker dissatisfaction. All three of these reactions have been linked to increased rates of accidents in other industrial settings.

It would be reasonable to expect accident rates to fall during the summer months when the weather improves, but they have either stayed the same or increased. During the summer months many experienced fishers leave the deep sea to fish in the coastal and midshore fisheries, and in August they traditionally take their vacations. This seasonal exodus makes it difficult to maintain good crews in the summer. This lack of experience – both on deck and in the wheelhouse – may be a major contributor to summer accidents. Here's how one captain described the effect of greenhorns on his boat:

Our steady bunch – everybody knew what to do. I'll give you an example. We took a new guy out two or three trips ago. We was haulin' back maybe twenty thousand ton of fish. The column [the bag of fish] was laying in over the stern. We were just puttin' a big choker on it to take it in, the guy was on the port side, and he wanted to go across the starboard side. Instead of walking around or going underneath the column he stood on it. When the ship made lift on a sea, it threw him right in the air. He landed on the deck which was a near miss, right. I said, "Why did you stand on that tow end to walk across, you know that's dangerous?" Then he said, "Well, I didn't think the ship would move while I was walkin' on it. I thought I could do it." But that is like playing Russian roulette. You don't do that with the sea. One of my steady crew wouldn't have done that. That was his first trip. But he was fishin' before. You know what I done then? Your blood stops circulating for about two minutes before you get back to life when you see somebody do that. Then you get him up and tell him.

Like I said earlier, you don't think the man could see what could happen. The tow end was laying there, he'd walk on it, not realizing that half a second from now that could be twenty feet off the deck. Well you're out to sea, and the boat moves, that's what you run into. That's what drives the skippers crazy. I know, myself, it gets to you after a while. You go out and you train people, and after two or three trips, well, God, you get a pretty good crew again. All of sudden, somebody else comes up. They got a job on a small boat, and they're gone. So you're right back where you started again. When you're pickin' up different crew every trip, you only get fellows who stay long enough to get a couple of bottles of wine. We don't want people like that. I really don't want that. If you haven't got people who's interested in doing it for a career, then it's no good to you. The money is not there to attract the good people into it. (Well Workers Interviews, Captain)

Frustration with these new working conditions and low monetary gains has led to an exodus of experienced crew members (such as certified trawlermen and scallopers) and officers, particularly captains and mates, who move from the deep sea into the midshore or the coastal fisheries. Some never return to the deep sea fishery. Others return only during the winter when other work is scarce.

This departure of captains and mates reduces the level of experience in the wheel-house. Many companies have tried to fill the gap by hiring well-trained and experienced workers from other deep sea companies, but few experienced deep sea captains and mates remain available. Companies also encourage young workers to attend fisheries school to receive or to upgrade their tickets, and sometimes the company pays them while they do so. But these workers lack experience and must rely on older and more experienced crew members or the company for advice. Inexperienced officers in the wheel-house can result in less efficient fishing when they or the crew use technology incorrectly or with difficulty, when they or the crew cannot repair equipment, or when accidents cause delays. Inexperienced officers can endanger the vessel, the crew, and the catch.

A similar problem can occur on the deck. For example, a certified trawlerman, after acquiring 270 sea days – about the average for one year's experience – can go to fisheries school and get the Mariner Third Class Certificate ("bosun ticket"). This presents problems, as one captain describes:

What they learn at school is not geared to running the deck but to working in the wheel-house. It's all right for them to know the bridge. They can know it inside out. But what I want on that deck is someone who can mend the nets, run the men, and knows how shooting and hauling works. It doesn't help me to have someone on the bridge. I can do that. (Well Workers Interviews, Captain)

An inexperienced bosun with greenhorns in the crew creates a dangerous combination. The bosun ends up relying on older hands without "tickets" to run the deck. While these more experienced workers can and do run the deck, they do not get paid for the extra responsibility because they lack a "ticket." If they get disgruntled enough to leave the vessel, it results in more greenhorns on deck and fewer "old hands" to fall back on. Reliance on untrained deckhands also increases the workload of experienced workers, who must teach and supervise the new workers as well as getting their own work done.

TO STAY OR TO GO?

This physically demanding and stressful environment, with long hours in harsh working conditions, grinds down men's bodies.[5] We do not know how many men leave the fishery annually. Many fishers simply "wear out" – drop out when they can no longer sustain the daily assault. As one veteran with forty years' experience states,

It always seemed to be a young man's game. Now the older fellows, they just seem to drift away. They just end up with bad backs in a lot of cases; in other cases, they're just not hardy enough to stay in the conditions. Most of the dropouts are just that. You've got to be in your prime years to stand up to this type of work. In most land-based industry, as you get older with your experience and your knowledge gained you'll probably step up the ladder and, generally speaking, it isn't a more rigorous position. You're able to use your mind and not as much of your body. But, I'm darn sure that a person of twenty-one who works for sixteen hours and rests for four comes back a lot quicker than someone even thirty-five.

I'd say for every one man that's fifty years old who's still scalloping, there'd have to be fifty who started with that man and dropped out through the years, through attrition. They just couldn't keep their end up. There's no way that on the share basis or the lay that they're going to carry their weight. The skipper's not going to carry a bunch of old men. When he gets the scallops on deck he's going to expect a lot of lively people running out there and getting 'em shucked out. The crew, as cruel as it may seem, makes life miserable enough for anyone who can't keep their end up. They'll get the message and they're not there next spring.

It's a contest every day. That's the name of the game in fishing. It don't make any difference what kind you once were, you're out. It's the only thing that's foremost in your mind. That's what keeps people working. (Well Workers Interviews, Mate)

Other fishers leave because of injuries, or because of employment opportunities in fish plants or in other fisheries, usually the coastal fishery. While records indicate how many fishermen hold licences, we do not know how many of these licences remain active. Many fishermen who leave to try other work retain their licences so that they will have the option of returning to the fishery. Even injured workers who return for a short time before dropping out remain officially "on the books." The fisher with chronic back problems simply decides not to go fishing any more. The records of the Nova Scotia Workers' Compensation Board indicate how many fishers received partial or full disability awards. These cases usually represent workers who had accidents the previous year or even earlier.

Our study suggests that for uninjured workers who leave, changes in working conditions associated with scientific management erode job satisfaction and upset the equilibrium between monetary and nonmonetary benefits associated with the job. Separation from family and friends, high stress levels, and erosion of autonomy remain major causes of job dissatisfaction. Crew most often cited those as reasons for leaving the fishery, while captains and officers cited the loss of autonomy and economic incentives in other fisheries as the most common reasons for leaving. Thus the level of stress combined with economic incentives in other sectors of the fishery push those fishers out of the deep sea as inevitably as injured workers leave.

Moving to the Inshore

Most captains who had left said they did so because they had become disillusioned with the deep sea fishing industry. Captains take on the job of site managers, as well as fish hunters and navigators. Onshore

managers now tell captains where to fish, what species to catch, and how much to bring back. Captains who leave see the company's control over the operation at sea as an infringement of their rights as captain.[6] As one captain who had worked the deep sea for twenty years put it,

The company I worked for didn't appreciate me ... There should be no boat quota. The fun was taken out of it. There were no highliners. It was just like going shopping at Sobey's. So I just quit. (Leavers Interviews, Captain)

Captains also see the way their earnings erode while the company increases its profits as a major reason for leaving the deep sea fishery. Combined with higher prices being offered in the inshore, this loss of earnings makes the coastal fishery more appealing. One captain explained the economic incentives for leaving this way:

I tell you why we're losing a lot of people now with our outfit. It's because of the inshore. The smaller boats are making a lot more money. Our people signed a four-year contract with the Company in '84 and it doesn't get over till the end of '88. You know what the market is done in fish prices. It's skyrocketed. We're making no money at all compared to what we used to make. Our prices is way down.

We're finding it so different because we're carrying different people all the time. If you had talked to me about two years ago, I wouldn't have told you about training because we didn't need it then. I can tell you out of Lunenburg alone we've lost more people in the last year than in the last ten – our good people too. So as far as we're concerned, the Company may be making a lot of money right now but they're losing a lot. When you lose your main resources, which is people, you're losing a lot, ain't ya?

They're payin' us really low prices. I know the contract is signed and everything, but why when they see what's going on, they don't give a little extra? They're making the money. The people that put the Company back on track is not getting paid for it. They want quality, quality, quality. How can you do good quality when you haven't got the people to do it? I can't do it alone. They've been told enough. There have been two or three of us. They just look at you, like you got two heads sometimes. You take a good man who's leaving, after that long, it makes you stop and wonder, doesn't it? He says, "I don't want to leave but I can make a lot more money somewhere else." You can't blame the man, can you? (Leavers Interviews, Captain)

The fishery oscillates between boom and bust years. Union contracts, which usually span three to five years, set the prices for that period without knowing what fish will be available in a given year. In some years and with some species the company wins; in other years or with other species the fishers do. In 1985 and 1986 the fishery expanded after a number of lean years. This fluctuation creates an economic incentive to join or to leave the fishery and has important effects on the crews of deep sea fishing vessels. As one captain explained,

The Company was sending out videos and showing our crews how much money they were making and all this. A lot of them says, "I really don't want to see this no more." They come to me and says, "Why should we watch this and we're not getting paid for what we're doing and the company's making millions?" So I went to the manager and I told him, "How do you tell a man to come in and watch a video showing the company making millions and some of those guys can't make their house payments?" He says, "Yeah, well I guess you're right." In 1981, deckhands on our boat made around fifty thousand dollars. That was a lot of money back then. I'm not saying that but you worked 280 days at sea. I think you deserve that much. But I can guarantee you this year, if they make thirty thousand, they're going to be lucky. So, I mean, how many other people in the last say seven years took a decrease in their pay that much? They said, "Oh well, that was a lot of money back then." But when people are used to making that money, and then take a cut like that, they're not going to stand for it. If they can come ashore and go on a small boat and make the money they were making seven years ago, they're going to do it, ain't they. I mean the inshore people is not catching so much fish as they were years ago, but they're making twice the money and they're getting paid for it. They're getting forty-five or forty-seven cents a pound for their codfish. And that's little trash. I mean they wouldn't take it from us. We had some codfish and we were lucky if we average seventeen or eighteen cents, and they're getting forty-three. It bothers you. I mean a man got to go out and catch four hundred thousand fish, and you're not making a living. It doesn't make sense. Well, our increases for this year was something like a quarter of a cent a pound. I don't have to tell you what the markets went up. It doesn't make sense, does it?

I mean, if we get say twenty cents a pound for fish, the company takes sixty-three per cent of that right off the top. We only get thirty-seven per cent of that. The boat takes sixty-three per cent and the crew gets thirty-seven per cent. I know one boat last trip there, the skipper lost a man that was with him seven years. He's gone longlining. He says, "Well, I really didn't want to go but I can make more money longlining and have more time home, so I'm gone." That's not only in Nova Scotia, that's in Newfoundland, too. All over. Well I wouldn't say all over but here and in Newfoundland, it's the same way. Their men feel the same way as we do, they're not being paid. (Well Workers Interviews, Captain)

These economic incentives mean that qualified trained fishers leave the deep sea fishery and begin fishing in the coastal fishery and they continue to do so while that fishery remains lucrative. Once the inshore fishery becomes less profitable than the offshore, those fishers return to the deep sea fishery. Thus the cyclic nature of the fishery translates into economic cycles mirrored by employment cycles. As the same captain continued to explain,

I was talking to Newfoundland skippers this trip, and one guy says last year they had something like two hundred names on a waiting list looking for jobs in Newfoundland on the trawlers. He says now they don't have anybody. There's not a name. One guy had five greenhorns, never was out before on one trip. Last year, you couldn't buy a chance in St. John's, Newfoundland; you couldn't buy a chance on them boats. So, there's the difference.

The first part of last year you couldn't buy a chance. Now with the ways things have gone, people are doing other things. I know my father fished in Newfoundland. He's got a small longliner. They're getting fifty cents a pound straight for codfish. I'm not saying they're making any fortune, mind you, but by Jesus they're making a living and they don't need much fish. He says they told him the other day that after Christmas they'll be getting sixty cents a pound. Here we are fishing with 1984 prices – other than a quarter or half a cent a pound increase. I mean people's not dumb anymore. Twenty years ago fishermen were – I wouldn't say stupid – but a lot of them was under-educated. They didn't read a newspaper. They came home and a lot of them got drunk and that's all that's to it. The situation today is changed. There is a lot of educated people at it. They know more of what's going on in the world than people gives them credit for sometimes. (Well Workers Interviews, Captain)

When captains or mates leave the deep sea they either buy their own vessel or work for a small independent owner on a vessel where the captain enjoys final control over when, where, and with whom he fishes. The trips have become significantly shorter and the time between trips substantially longer. These smaller vessels have cramped living quarters, smaller work decks, and less safety equipment. With no large company to run interference, captains must deal directly with federal and provincial fisheries officers. Workers trade off those less pleasant working conditions, for shorter sea time, increased flexibility in working conditions, and increased autonomy.

The "other" officer group presents a more complicated picture. These men have left the fishery because of injury, physical stress, or stifled ambition. The majority do not possess sufficient resources to become owner/operators, but most hope to upgrade to mate or captain. They suffer a loss of status. They must now direct fewer workers on a smaller deck or in cramped fishrooms. They must bunk with the rest of the crew. Their higher level of job satisfaction with deep sea living conditions and the general physical environment can be traced to their current working environment. These fishers trade off the disadvantage of less pleasant working conditions for the advantages of less time at sea, more autonomy and input into their working conditions, and better possibilities of advancement.[7]

The coastal fishery appears attractive in other ways. Its organization more closely resembles the craft enterprise – more personal and less bureaucratic. Here the owner/operator or captain directly controls the enterprise. The sense of challenge and adventure remains acute. The captain recruits his own crew based on their expertise as fishers. The captain/owner pays the crew on a share basis after he has deducted the expenses of the trip and of the boat, including such costs as mortgage and loan payments, from the gross. All take the risks; all reap the benefits or share the losses.

In general the organization of work in the coastal fishery appears more sensitive to fishers' needs. Time at sea depends on how long it takes to get the catch and bring it back to port and time on shore has expanded between trips. (This tendency has increased under the ITQ system, where each vessel has its own quota and must stop when it is reached.) A more informal work schedule makes it easier to take off a trip or to postpone sailing. The crew develops a strong sense of identity and group solidarity through common experiences, goals, and shared values. It gives back to deep sea workers a sense of personal control of their working conditions.

Workers in the inshore fishery do not assume its security or future. Traditional values remain strong and closely tied to job satisfaction – trip times, personal responsibility, small tightly knit crews, challenge, freedom to choose, "high" earnings for the region. Rising costs of vessels and gear, falling quotas because of exhaustion of stocks, increased efforts of vessels from other sectors of the distant water fleet, and greater harvesting capacity all indicate economic uncertainties ahead. So once again nonmonetary rewards must be balanced against monetary ones. Once the balance tips in favour of the deep sea, some former deep sea fishers will leave the coastal fishery and return to the offshore fishery.

In fundamental ways, workers' explanations remain incomplete and unsatisfying. Because fishers give their reasons for leaving in personal terms, their explanations tend to mask the way in which corporate management practices in the fisheries have imposed a work regimen that in itself causes increased safety risks, stress, and job dissatisfaction. The workers sometimes fail to note that their dissatisfaction stems not always from the nature of the work but sometimes from the results of management techniques. The size of enterprise allocation, the fluctuation of market demands, the needs of the fish processing plants, and the introduction of new fishing technology all conspire to introduce discordant rhythms.

CHAPTER SEVEN

Counting the Dangers

We do not know how many men are killed or injured annually in the deep sea fishery. Statistics generated by government agencies such as the Nova Scotia Workers' Compensation Board or Marine Casualty Investigations have inherent biases that only allow us to estimate morbidity and mortality rates.[1] In this chapter I describe each of these agencies and their mandates, then discuss fishers' morbidity and mortality patterns generated from the agencies' data and additional material derived from my 1986 and 1987 surveys.

Both employers' and workers' costs can be measured using statistics and reports from state agencies. Workers' costs can be measured in terms of deaths, injuries, and lost earnings. Employers' costs for physical risks include insurance premium payments, liability settlements, down time, the cost of replacing workers, and new equipment costs. Both labour and capital rely on government statistics to support their claims in labour conflicts, because statistics produced by governments generally appear "objective" and "scientific." Reports from people working in the industry – either labour or capital – reflect particular interests and therefore appear to be very "subjective." When labour collects its own data, employers and buyers always receive the results with suspicion. The statistics generated by government agencies such as Marine Casualty Investigations and Workers' Compensation Boards allow labour to appeal to employers – through contract negotiations – and to government by basing their demands on supposedly objective and scientific evidence.

However, government statistics also have biases. First, those statistics reflect the mandate of specific government agencies and their understanding of the conflict between labour and capital. Second, they indicate the government's position as mediator between labour and capital. Third, governments collect data for reasons other than simply documenting rates of injuries and deaths. The lack of certifiable work-related statistics impedes the state's recognition of workers' needs in the area of workers' health and safety.

THE NOVA SCOTIA WORKERS' COMPENSATION BOARD

Under the Nova Scotia *Workers' Compensation Act*, as revised in July 1968, the board provides compensation to all disabled or injured workers and aids in accident prevention. It offers, in principle, a no-fault insurance system whereby companies pay premiums to the board and workers receive benefits if injured on the job. The board examines an incident in order to ascertain if an injury or illness originated from working conditions; if it did, the board must then assess the level of disability and determine the duration of compensation. The board does not investigate the question of liability, and, in order to get the benefits, workers must give up the right to sue their employers. In 1971 the Workers' Compensation Board began to include deep sea fishers under the provisions of the 1968 Act.

The Nova Scotia *Workers' Compensation Act* excludes from benefits and premiums workers in every industry in companies where fewer than three workers work at the same time. Fishing vessels with crews of three or more – excluding the skipper – meet the eligibility requirements of the board. An individual who does not qualify can be voluntarily self-insured, but because of the cost few fishers do so. Out of 12,560 licensed fishers in Nova Scotia in 1984, only 5,960, or 47.5 percent, worked on vessels with crews of three or more (Canada, Department of Fisheries and Oceans 1984).

The board's restriction on crew size roughly corresponds to the Department of Fisheries and Oceans' classification of forty-five feet and over for boats working the offshore. Those boats include trawlers, draggers, scallop draggers, and seiners. This limitation of eligibility not only decreases the number of claims but also restricts the type of claims. For example, lobster boats seldom have a crew of three. The gear on those boats differs substantially from equipment on larger vessels, and injuries vary correspondingly; gear seldom crushes lobster workers, but they often fall overboard when hauling pots.

Workers can receive three types of compensation: total disability, partial disability, and medical aid. If a work-related accident totally disables a worker for at least three days, he/she can claim "compensation computed at the rate of seventy-five per cent of his [*sic*] average yearly earnings paid from the date of the accident and paid for the period of disability" (Nova Scotia, Department of Labour 1985, 10).

In 1985 the board paid a maximum of $28,000 per year and a minimum of $120 per week (or $6,240 per year). If a worker becomes "partially disabled the compensation is seventy-five per cent of the difference between the average earnings before the accident and the average amount which the employee is able to earn after the accident, and such compensation is payable only so long as the disability lasts" (ibid. 1985, 10). Medical aid includes hospitalization, medical care, surgery, medical and physical rehabilitation, prescribed drugs, and other medical expenses arising from a work-related injury. Workers with permanent disabilities, either total or partial, receive compensation for life. In 1985 the board paid a maximum of $28,000 per year but a minimum of $530 per month. If a work-related injury results in death the board pays modest burial expenses and transport costs for the return of the body. Widows/widowers receive a lump sum ($1,000 in 1985) and a monthly payment ($569 in 1985) until their death or remarriage (Nova Scotia, Department of Labour 1985, 10). Dependent children receive a monthly payment ($149 in 1985) until the age of eighteen or, if they remain in school full time, until twenty-one. If a worker has no spouse or children but has other dependents, such as a mother or father, they receive compensation.

Figure 7.1 illustrates the process of filing a claim. The first stage consists of the employee informing the employer and a doctor of the accident, and, once all the necessary forms have been filled out, the employer notifying the Workers' Compensation Board of the accident. This initial stage takes about twenty days. Next the board processes the claim and collects any missing information. Once the file has been completed and summarized, it goes forward for adjudication. This second stage usually takes over a month. Once a decision on the claim has been made the board informs the worker and sends a cheque. If the employee wishes to appeal the decision he/she must launch a formal complaint with the board immediately. Since appeals consume time and money, few employees want to go through this adversarial process.

The provincial governments regulate Workers' Compensation Boards and define their mandates. Because governments have taken on this responsibility, the costs generated by workplace hazards have changed into social costs borne by the employees and by the community at large rather than by capital. Employers pay premiums to the board for the cost of administering the plan. Thus governments and employers portray compensation as an insurance benefit, rather than a right to reimbursement for health and income loss sustained during the course of work. The boards relieve employers of the primary responsibility for work-related illnesses and injuries and of the burden

Days elapsed	Employee	Employer	WCB Agent
0	Accident occurs Notifies doctor and employer	Receives notification	
	Completes forms	Completes forms	
20		Forwards forms and medical reports	Receives notification
			To claims processor Reviews files and forms Requests missing information
	Notified of missing information Forwards missing information	Notified of missing information Forwards missing information	
			Receives missing information
56			Summarizes claim file
58			Claim to adjudicator
67			Decision made Payment sent
	Notified of decision Launches appeal (if necessary)	Notified of decision	

Figure 7.1
Processing a Workers' Compensation Board claim
Source: Nova Scotia, Department of Labour, *A Measure of Performance: Annual Report of the Workers' Compensation Board of Nova Scotia* (1990, 4–5).

of potentially huge cash awards granted by judges or juries in liability trials.

Governments also fund compensation boards on the premise that accidents constitute a predictable part of employment, and they assume that workers principally cause accidents. Active trade union health and safety campaigns have made more workers aware of this erroneous assumption. Governments and employers must take responsibility for instructing both labour and management about health and safety hazards in the workplace. Yet the onus remains on workers to show how their illnesses or injuries relate to their work, and to prove the level of their disability. Many workers do not report cases because of the slowness and the adversarial nature of the process. Many more cases go unreported because workers see them as routine injuries that

Table 7.1
Fishers' accidents for 1978–88 by year and type of injury

| | | Type of compensation | | | | |
Year	Fatal	Permanently disabled	Temporarily disabled	Medical aid only	Non-compensable	Total
1988	0	44	356	149	32	581
1987	1	0	283	133	35	452
1986	8	0	342	136	27	513
1985	1	1	295	130	19	446
1984	1	0	312	129	23	465
1983	1	1	341	142	16	501
1982	4	1	332	148	22	507
1981	2	1	340	179	25	547
1980	6	0	308	144	15	474
1979	6	1	264	117	15	403
1978	2	5	272	118	21	418

Source: Compiled from Nova Scotia, *Annual Reports of the Workers' (Workmen's) Compensation Board of Nova Scotia,* 1978–88, Table 1, pp. 48, 48, 50, 51, 49, 49, 49, 20, 23, 23, and 23 respectively.

do not warrant the loss of wages. For example, Nova Scotia deep sea fishers interviewed in the spring of 1986 felt that things such as cuts, bruises, hand and finger injuries, pulled muscles, back injuries, and fish-hooks caught in hands and legs were routine, and thought it unlikely that they would seek compensation for them.

The board reports annually on the types of injuries and illnesses compensated for, on the level of disability, and on the number of claims made. Table 7.1 indicates the type of compensation fishers received for their injuries from 1978 to 1988. Note that the Workers' Compensation Board of Nova Scotia defines the unit of analysis as the claim, not as the individual making the claim. Since an individual can make more than one claim in a year, the number of claims can be higher than the number of people being compensated in any given year.

Tables 7.2 and 7.3 summarize the causes of accidents and the nature of injuries sustained by fishers who settled with the Nova Scotia Workers' Compensation Board between 1978 and 1985. In 1986 the board changed the format of their reports; from 1986 on the annual reports contained less information on the characteristics of accidents and injuries sustained and on compensation given (e.g., the board did not report accidents and injuries by individual industries). In these later reports the board highlighted the financial responsibility of the fund (i.e., the financial self-sufficiency of the board and its limited use of

Table 7.2
Compensable cases settled by the Nova Scotia Workers' Compensation Board
between 1978 and 1985 by cause of accident, for class 2*

Cause of accident	1978	1979	1980	1981	1982	1983	1984	1985
Striking against	15	24	30	34	31	43	27	27
Struck by	66	93	73	126	110	102	93	86
Caught in, on, or between	39	36	34	44	35	33	32	42
Collisions, derailments, and wrecks	2	1	6	3	1	2	2	2
Falls and slips	106	81	60	103	109	101	115	104
Conflagration, temperature extremes, and explosions	1	5	5	3	10	4	5	6
Inhalations, contact, absorptions, ingestions, industrial diseases	3	3	2	1	1	1	2	1
Contact with electrical current	0	0	0	0	0	0	0	0
Over-exertion	40	38	47	82	92	102	97	89
Miscellaneous accident types	5	2	4	4	8	4	3	2
Totals	275	283	261	400	396	392	377	352

Source: Compiled from Nova Scotia, *Annual Reports of the Workers' (Workmen's) Compensation Board of Nova Scotia*, 1978–85, Table 6, pp. 54, 55, 56, 58, 58, 56, 56, and 56 respectively.
*Fishers.

Table 7.3
Compensable cases settled by the Nova Scotia Workers' Compensation Board
between 1978 and 1985 by nature of injury, for class 2*

Nature of injury	1978	1979	1980	1981	1982	1983	1984	1985
Amputations	4	4	1	8	4	2	6	3
Burns and scalds	0	4	6	4	8	5	7	7
Cuts, lacerations, and punctures	119	117	98	150	148	141	127	106
Sprains and strains	81	66	86	141	52	160	143	150
Fractures and dislocations	35	43	35	60	52	44	55	50
Eye injuries	12	11	2	4	3	4	4	4
Herniae	0	0	0	0	0	0	0	89
Industrial diseases	7	11	2	4	3	4	1	4
All other injuries	16	15	18	16	19	32	26	21
Totals	275	283	261	400	396	392	377	352

Source: Compiled from Nova Scotia, *Annual Reports of the Workers' (Workmen's) Compensation Board of Nova Scotia*, 1978–85, Table 4, pp. 52, 53, 54, 56, 56, 54, 54, and 54, respectively.
*Fishers.

public funds), accident prevention programs, and public service. This change in format more honestly reflects the mandate of the board. (Note: Fishers filed more claims for compensation from the Nova Scotia board than Marine Casualty Investigations identified for all of Canada.)

The Workers' Compensation Board's characterization of accidents and injuries reflects their mandate: accident prevention and compensation. The board lists the cause of the accident and the nature of the injury. The injuries listed include amputations; burns and scalds; cuts, lacerations, and punctures; sprains and strains; fractures and dislocations; eye injuries; herniae; industrial diseases; and all other injuries (see table 7.3). Except for industrial diseases, all arise from traumatic situations and can be linked to a specific event. The agency records a few cases of compensation for chronic ailments in the single category of "industrial diseases." The *Workers' Compensation Act* of Nova Scotia recognizes about fifteen diseases as work-related. Although no specific disease has been directly linked to fishing, cases recorded under the category "industrial diseases" include chronic ailments attributed to this kind of work. For example, doctors have begun to recognize back ailments as a chronic condition for fishers, but they have not yet been certified as an industrial disease. The difficulty lies in the definition of an industrial disease as a disease or condition that a worker can get *only* by working in that particular industry, and in proving that a condition of a general nature, such as lower back pain, relates to work on a trawler and not to other activities outside the workplace.

MARINE CASUALTY INVESTIGATIONS

Marine Casualty Investigations, part of the Department of Transport, investigates all marine casualties (mishaps to vessels) and accidents, including those on merchant vessels, fishing boats, and oil rigs, in order to determine whether incidents involve damage to the vessel, to define the cause, and, most important, to assess legal responsibility. In 1983 the government licensed about 12,560 fishers in Nova Scotia (Canada, Department of Fisheries and Oceans 1984). For fishers, the agency includes all accidents aboard ship and all marine casualties involving fishing vessels – from company trawlers to small dories. Indirectly then, they define all Nova Scotian fishers, coastal or deep sea, as at risk.

Evidence gathered by Marine Casualty Investigations can be used in any later court action brought by the parties involved. As part of the Department of Transport (Transport Canada), it recommends modifications in or initiates marine safety legislation and helps to enforce those regulations. When a vessel gets into trouble it provides services for the vessel and its personnel through Search and Rescue. Marine Casualty Investigations reports on these incidents annually, and on any resulting deaths and injuries. It does not directly investigate fatalities or injuries, nor does it assess health risks.

Marine Casualty Investigations divides incidents into two categories, casualties and accidents aboard ship. According to the agency, two types of casualties exist. Both refer to the vessel, and not the personnel: (1) "collision, grounding, contacting, striking, foundering, sinking, fire, explosion, capsizing, and ice damage" and (2) "any other type of accident in which a vessel has been damaged" (Canada, Transport Canada 1982–91, Preface to 1983).

This agency defines an accident aboard ship negatively, as "an accident on board a vessel resulting in death or injury not resulting from a marine casualty" (ibid.). A simple example will make the distinction clear. If worker on a trawler slips, falls, and breaks a leg while going down a ladder, the incident would generally be classified as an accident aboard ship. However, if the vessel collided with another vessel, causing a worker going down a ladder to slip, fall, and break his leg, that would be classified as a marine casualty.

The agency distinguishes between the two types of incidents by its legal mandate. Marine Casualty usually views "accidents aboard ship" as straightforward, since they do not require a thorough investigation and seldom result in legal action; the worker who slipped when going down a ladder has had a routine accident independent of vessel damage. The need to report and investigate marine casualties arises from the effort to mediate between interests of owners, and to designate the party at fault so that liability can be assigned. It does not depend on a desire to reduce or eliminate physical risks in the work place per se.

It is only since the late 1960s that Marine Casualty Investigations has recorded both deaths and injuries aboard ship. Before this the agency only reported mishaps to vessels. The inclusion of the category "accidents aboard ship" recognizes work-related incidents. Although the underlying assumption that workers' compensation will look after workers who suffer routine injuries remains, the agency investigates fatalities in order to ascertain responsibility and mediate between the interests of the holders of capital.

This mandate also reflects the extent of the reporting of the different types of accidents investigated: the more severe and extraordinary the injury, the more thorough the analysis. Marine Casualty Investigations claims that it identified and investigated all *fatalities* associated with marine casualties or accidents aboard ship. The agency estimates that its investigations include only 20 percent of the Workers' Compensation Board's cases of *injured* workers.[2] The reporting of injuries to workers also reflects the limited money available for investigations and the relatively low priority given to fishers' safety concerns by the government.

Three levels of investigation exist. For all reported incidents, a formal precoded report[3] must be filled out by the master of the vessel or other appropriate person (for example, a management representative). The unit of analysis remains the vessel and not the person or persons who work on the it – a characterization that stresses owner/employer interests and property investments over workers' safety. The form reflects the mandate of the agency by stressing first the causes of marine casualty and then the nature of injuries caused by either a marine casualty or an accident aboard ship. About two-thirds of all investigations end at this level, because the agency can identify a clear cause for the accident and sees no need for legal action. For the remaining cases a second level of investigation takes place. Local Marine Casualty Investigations officers interview the major actors. If no legal action appears appropriate, the investigator concludes the investigation and makes a public report. Only about one-twelfth of all the cases go beyond that stage to a formal inquiry. The evidence gathered during the inquiry is taken under oath and can be used in a court of law, but those reports remain off the public record. Usually one or both parties involved will follow up with legal action. Recommendations from these inquiries influence safety standards (for example, for master recertification, use of government resources such as navigational aids, and tackle regulations).

Owners recognize the importance of this development, and at least one company has established a safety program to train its captains and mates about safety hazards on board. The company also makes it clear to those people that they will be held liable for errors in judgment if a marine casualty or an accident aboard ship occurs.

The numbers of deaths and injuries for 1981–89 resulting from marine casualties or accidents aboard ship involving fishing vessels – as compiled by Marine Casualty Investigations – appear in table 7.4. Tables 7.5 through 7.8 reflect the separation of the agency's tasks. Tables 7.5 and 7.6 include all major types of marine casualties listed in its definition. Tables 7.7 and 7.8 divide into two segments: the type of accident (or incident) and the cause. "Accident" categories include asphyxiated, burned, crushed, drowned, electrocuted, exhaustion, injury, lost ashore, malicious action, malnutrition,[4] missing, poisoned, shock, unconscious, and "other." These categories include a wide

Table 7.4
Marine occurrences (pleasure craft excluded) for Canadian fishing vessels, 1981–90

	Deaths			Injuries		
	Marine casualties	Accidents aboard ship	Total	Marine casualties	Accidents aboard ship	Total
1981	15	10	25	7	50	57
1982	10	16	26	11	55	66
1983	15	8	23	23	67	90
1984	26	7	33	20	39	59
1985	30	9	39	27	66	93
1986	19	4	23	25	79	104
1987	38	12	50	27	101	128
1988	24	12	36	33	106	136
1989	18	16	34	26	111	137
1990*	25	11	36	24	77	42

Source: Canada, Transport Canada, *Statistical Summary of Marine Accidents* (1982–84) for 1981–83, Table 5, pp. 8–9; (1985–90) for 1984–89, Tables 3, 4, pp. 4–5; Canada, Transportation Safety Board, *Annual Report,* 1990 (Ottawa, 1991), Tables A-3, A-4, p. 101.
*From 1981 to 1989 the summary statistics were published in two places: the *Statistical Summary of Marine Accidents* compiled by Marine Casualty Investigations, and the *Annual Report* of the Transportation Safety Board. There are slight discrepancies. After 1989, they were published only in the *Annual Report* of the Transportation Safety Board. I have used the Transportation Safety Board statistics only for 1990.

Table 7.5
Deaths as a result of marine casualties involving fishing vessels, 1981–89

Casualty	1981	1982	1983	1984	1985	1986	1987	1988	1989
Collision	0	0	0	0	0	3	0	0	0
Grounding	3	0	3	3	0	1	2	2	0
Striking	0	0	0	1	2	0	0	0	0
Contact	0	0	0	0	0	0	0	0	0
Foundering	4	0	0	8	10	2	13	4	3
Sinking	0	5	0	0	0	5	1	1	2
Fire	5	1	0	2	2	0	1	0	0
Explosion	0	0	0	0	0	0	2	0	1
Capsizing	3	4	11	8	9	7	3	9	3
Ice damage	0	0	0	0	0	n/a	n/a	n/a	n/a
Other	0	0	1	4	7	1	16	8	9
Totals	15	10	15	26	30	19	38	24	18

Source: Compiled from Canada, Transport Canada, *Statistical Summary of Marine Accidents* (1982, 1983) for 1981, 1982, Tables 3 and 6, pp. 8, 12 and 6, 10; (1984–90) for 1983–89, Table 8, p. 9. n/a = not available.

Table 7.6
Injuries as a result of marine casualties involving fishing vessels, 1981–89

Casualty	1981	1982	1983	1984	1985	1986	1987	1988	1989
Collision	1	4	1	0	1	3	0	2	1
Grounding	0	3	3	4	7	8	0	8	5
Striking	0	0	0	0	1	0	0	1	0
Contact	0	0	0	0	0	0	0	1	0
Foundering	0	0	0	0	4	1	2	0	1
Sinking	1	0	0	2	0	0	0	0	2
Fire	0	0	2	8	4	3	2	6	2
Explosion	3	3	11	5	5	6	15	11	15
Capsizing	0	1	2	1	2	n/a	n/a	n/a	n/a
Ice damage	0	0	0	0	0	2	5	0	0
Other	0	0	4	0	3	2	3	4	0
Totals	7	11	23	20	27	25	27	33	26

Source: Compiled from Canada, Transport Canada, *Statistical Summary of Marine Accidents* (1982, 1983) for 1981, 1982, Tables 3 and 6, pp. 8, 12 and 6, 10; (1984–90) for 1983–89, Table 8, p. 9.
n/a = not available.

range of cases, but do not offer much information. They reflect a primary interest in collecting data on causes of death rather than types of injuries. Except for malnutrition, the accident categories originate from traumatic situations directly linked to the immediate environment. During the investigation process, the agency makes no attempt to include information outside of this environment. For example, they do not consider the stress related to speed-up during heavy fishing.

The categories for type of incident include carried overboard, caught in machinery, exposure, fell overboard, fell in to holds/tanks, fell on deck, fight, heavy weather, suicide, and "other." Both sets of categories refer specifically to seagoing vessels. They also reflect an event-oriented approach, dealing with the type of accident and its cause, but not with the type of injury sustained. For example, the accident category is "crushed," rather than "amputation," or "death," which would be the result of the accident.

WHOSE DEFINITION IS
THE RIGHT ONE?

Just how many types of injuries are not recorded? Neither of these agencies reports cases of chronic injuries, or illnesses other than malnutrition. This characterization of injuries limits the potential inclusion of endemic and environmental conditions. For example, if

Table 7.7
Deaths not involving a vessel casualty associated with fishing vessels, 1981–89

	1981	1982	1983	1984	1985	1986	1987	1988	1989
Accidents									
Asphyxiated	1	0	0	1					
Burned	0	0	0	0					
Crushed	1	1	3	1					
Drowned	7	11	3	4					
Electrocuted	0	0	0	0					
Exhaustion	0	2	0	0					
Injury	1	2	2	0					
Lost ashore	0	0	0	0					
Malicious action	0	0	0	0					
Malnutrition	0	0	0	0					
Missing	0	0	0	0					
Poisoned	0	0	0	0					
Shock	0	0	0	0					
Unconscious	0	0	0	0					
Other	0	0	0	0					
Totals	10	16	8	6					
Incidents									
Carried overboard	1	0	0	0	1	0	2	2	0
Caught by machinery	2	3	4	2	1	0	0	1	2
Exposure	0	2	0	1	n/a	n/a	n/a	n/a	n/a
Fell overboard	4	9	2	3	7	4	7	5	9
Fell into holds/tanks	0	0	0	1	0	0	0	0	0
Fell on deck	0	0	1	0	0	0	0	0	0
Fight	0	0	0	0	n/a	n/a	n/a	n/a	n/a
Heavy weather	0	0	0	0	0	0	0	2	1
Suicide	1	0	0	0	0	0	0	1	0
Other	2	2	1	0	0	0	3	1	4
Totals	10	16	8	7	9	4	12	12	16

Source: Compiled from Canada, Transport Canada, *Statistical Summary of Marine Accidents* (1982–85) for 1981–84, Tables 2 and 3, pp. 6–7, 8–9; (1986–90) for 1985–89, Table 3, pp. 4–5.
Note: Information on accident type not available after 1984.
n/a = not available.

a worker becomes incapacitated because of chronic back pain caused by repeated "routine" falls on board a vessel, Marine Casualty Investigations does not consider him to be an accident victim. He cannot sue the company, because his case falls under the jurisdiction of the Workers' Compensation Board. A specific, exceptional event must be involved in causing the injury in order for the agencies to characterize the incident.

Table 7.8
Injuries not involving a vessel casualty associated with fishing vessels, 1981–89

	1981	1982	1983	1984	1985	1986	1987	1988	1989
Accidents									
Asphyxiated	0	0	3	3					
Burned	0	0	2	2					
Crushed	1	8	4	1					
Drowned	0	0	0	0					
Electrocuted	0	1	0	0					
Exhaustion	0	0	0	0					
Injury	48	45	58	31					
Lost ashore	0	0	0	0					
Malicious action	1	1	0	1					
Malnutrition	0	0	0	0					
Missing	0	0	0	0					
Poisoned	0	0	0	0					
Shock	0	0	0	0					
Unconscious	0	0	0	1					
Other	0	0	0	0					
Totals	50	55	67	39					
Incidents									
Carried overboard	0	0	3	0	0	0	0	0	0
Caught by machinery	27	29	38	26	43	53	45	52	40
Exposure	0	0	0	0	n/a	n/a	n/a	n/a	n/a
Fell overboard	2	1	1	1	2	1	2	4	5
Fell into holds/tanks	3	1	0	1	0	5	0	3	1
Fell on deck	6	12	13	4	8	5	19	14	12
Fight	0	0	0	0	0	n/a	n/a	n/a	n/a
Heavy weather	0	2	2	0	1	3	0	5	2
Suicide	0	0	0	0	0	0	0	0	1
Other	12	10	10	7	12	12	35	28	50
Totals	50	55	67	39	66	79	101	106	111

Source: Compiled from Canada, Transport Canada, *Statistical Summary of Marine Accidents* (1982–85) for 1981–84, Tables 4 and 7, pp. 10–11 and 12–13; (1986–90) for 1985–89, Table 7, pp. 8–9.
Note: Information on accident type not available after 1984.
n/a = not available.

Even when an event does occur, it may still be excluded. For example, if a worker suffers a heart attack and dies while working the trawl or the drags, Marine Casualty Investigations does not include his death in its statistics: the death does not fall into either of the categories of marine casualty or an accident aboard ship. The agency will argue that although this unfortunate death happened at sea, they cannot classify it as either of those and cannot include it in their statistics because it involved neither an outside agent nor liability. They assume death from natural causes and therefore it remains outside their mandate.

The definition of an incident as an "accident aboard ship" does not preclude its inclusion, but the agency's definition of the *category* "accident aboard ship" does. If the agency viewed such deaths as exceptional and not routine it would include them, but it does not. It sees them as an ordinary cost paid by labour.

All of the agencies structure their statistics by cause of accident, but their categories differ (see tables 7.2, 7.7, and 7.8). Because of the more general nature of the Workers' Compensation Board's mandate, which includes all industries, it defines its categories broadly and includes types of accidents not applicable to marine vessels (e.g., derailments). Marine Casualty Investigations's categories reflect its mandate. They relate directly to marine vessels (e.g., "fell overboard" or "carried overboard"). Marine Casualty Investigations gives much more detailed information about the type of accident than does the Workers' Compensation Board.

Both the government agencies have narrowly defined views of occupational health and safety concerns consistent with their mandates. As state agencies, Marine Casualty Investigations Division and the Workers' Compensation Board of Nova Scotia constitute integral parts of a system that controls the recording of vital statistics, social welfare benefits, and industrial regulations, and reflects the state's role as mediator between labour and capital interests in the fishing industry. These agencies define "accidents," "deaths," and "disabilities" in terms of those interests, and the categories reflect those concerns. As Smith (1975, 96–97) argues,

The bureaucratic and professional terminologies and methods of making accounts of these (various more or less formal kinds) are not inessential. They cannot be arbitrarily dispensed with. The terms and methods of accounting for and describing what is being done set the forms under which people recognize what they do and what others do as properly actionable ... The world that people live in and in which their troubles arise is "entered" into the system set up to control it by fitting them and their troubles to standardized terms and procedures under which they can be formally recognized and made actionable. These procedures are *intrinsic* to the workings of professional and bureaucratic forms of organization.

Thus defining a death or injury as work related does not simply give certain characteristics to an accident or casualty, but it certifies or makes real the relationship between the work and the accident or casualty. (And, conversely, not so defining a death or injury tends to deny that relationship.) The definition expresses a relationship between the state's interests, the established frame of reference for categorizing those interests, and an event in the practical activities regulated by the state. The characterization of injuries and accidents by the state agencies reflects the theoretical framework described above.

The definitions used by these agencies restrict the gathering of information on work-related illnesses and injuries. They focus their attention either on the workplace – the vessel, its equipment, its immediate environment – or on a few significant but negative health features such as drug and alcohol abuse or obesity. They attempt to compile an "unbiased," "scientific" set of statistics relating to incidents directly involving the work environment. This production of an objective set of data defines the immediate work environment for discussion. It also defines types of work-related accidents, injuries, and illnesses as unacceptable (included) or acceptable (excluded) for workers to endure. Labour, through unions and workers' groups, lobbies for the inclusion of categories or a redefinition of work-related incidents. Employers meanwhile attempt to maintain the status quo.

The debate now revolves around the identification of physical, chemical, and psychological agents that cause accidents or cause workers to become sick, and on the elimination of these agents from the work environment. The agencies discussed in this chapter believe that the immediate physical or social environment of work, the behaviour of workers, or an interaction of both causes accidents (Reschenthaler 1979, 24). Since they assume worker resposibility – wholly or partially – for accidents, they presume that controlling the work environment through education, training, regulation, and technological innovation will solve those problems. They pay little attention to the effects of social and economic constraints on the work experiences of crews aboard vessels. The model developed by these agencies divorces the vessel and the crew from their port and community.

This attitude pervades all aspects of their mandates, including training programs offered to workers and employers and recommendations for modifications in safety legislation. For example, when a worker fell out of his bunk in rough winter seas, Marine Casualty Investigations recommended devices to keep workers in their bunks during rough weather. It did not ask the question, "Why was the vessel out in such a gale?"[5]

The recognition of capital and its basic policies may have greater influence on safety than many of the recommendations proposed by the state. A growing body of literature (Bergman 1978; Reasons, Ross, and Paterson 1981; Sass 1979) identifies circumstances outside the workplace that have a profound effect on workers' occupational health and safety. Reasons, Ross, and Paterson (1981) point out that short-term economic fluctuations, structural alterations in industry, techno-logical adaptations (such as factory freezer trawlers), and social changes in workers' communities invariably affect working conditions. For example, if a company pays a captain a bonus for large quantities of fish, the captain may overload the vessel, putting his crew at risk, in order to get the premium. If the company changed its policy and offered a bonus for quality and not quantity of fish caught, then this risk would be reduced.

Many of the state's safety recommendations address only the super-ficial causes of incidents – and offer stopgap measures to deal with them – without attempting to explore the underlying causes. These agencies have little incentive to analyse their statistics in terms of the technical sophistication of the workplace or to see them as an expres-sion of specific social relations. Navarro (1982, 11) argues that "the physical effort needed to execute the work; the interaction between the workers, the objects of work, and the means of labour; and the degree of control that the workers have over the means of work and over the process of work" define areas of social relations that must be examined. Under specific conditions those circumstances can increase the likelihood of accidents. For example, the speed of a mechanical winch on a trawler imposes a specific work pace that the worker cannot alter. In an extreme case, a condition like that could lead to a heart attack because of the physical effort involved. By looking at the orga-nization of work, and the demands of technology on the worker, many other illnesses and conditions could be identified as work-related. For example, chronic stomach distress plagues many deep sea fishers. This condition results from a combination of work-related circumstances, including diet, action of the vessel, stress, and the work routine (which can involve vigorous exercise or sleep right after a meal). Yet the agency does not consider chronic stomach distress an occupational disease. In the fishery, as in other industries, more and more people now understand that circumstances other than those in the immediate work environment affect workers' health.

ACCIDENTS AND INJURIES

Although the data have inherent biases, we can learn a great deal about morbidity and mortality patterns among Nova Scotia deep sea fishers from the material gathered from Workers' Compensation Board and Marine Casualty Investigations. According to Workers' Compensation Board material the most common causes of accidents include slips and falls, over-exertion, and striking against or being struck by or caught in, on, or between machinery or gear (see Table 7.2). The board most frequently compensated injuries involving cuts, lacerations and punctures, sprains and strains, and fractures and dislocations. Although we can identify the frequency of types of injuries and the causes of the accidents from the board's material, we cannot correlate them or put them into context.

Since the Marine Casualty Investigations material distinguishes between marine casualties and accidents aboard ship, we can put deaths and injuries into context. According to the Marine Casualty Investigations material on marine casualties, deaths occurred most frequently with capsizing, foundering, sinkings, groundings and fires, while injuries occured most frequently with explosions, groundings, fires, and collisions (see tables 7.5 and 7.6). Tables 7.7 and 7.8 indicate the numbers of deaths and injuries associated with accidents aboard ship. The most common causes of death included drowning and being crushed as the result of falling or being carried overboard and being caught in machinery. The information on injuries is less helpful since "injury" itself is used as the most common category of accident; however, being caught by machinery, falls on deck and into holds and tanks, and falling overboard appear to be the most frequent causes of injuries not involving a vessel casualty. Thus material from each agency complements that of the other.

My own research did not allow me to calculate accident rates for fishers either; however, I could identify types of accidents and injuries and related patterns that could not be gleaned from the government agencies' material. From the 1986 survey I could identify 46 men who had suffered a serious accident during the previous year. Since the sample only included workers still engaged in fishing, I assumed that none of those serious accidents stopped them from working permanently in the fishery, but did result in lost work time.[6] The 1987 survey of workers who had left the fishery included 59 men who had left primarily for medical reasons. The accident patterns for those disabled workers differed from those of the "well workers" sample (see tables 7.9 and 7.10).

For discussion of accidents and injuries I categorized accidents as follows: bruising/blows, wounds, stabs, sprains and strains, fractures, crushing, burns, frostbite, embedded foreign bodies, and blood poisoning. Table 7.9 indicates the percentage of accidents to well workers by type and body part. Well workers most often suffer from cuts or crushed fingers. The most common types of accidents to all body parts included wounds (30.2%) and crushings (28.6%); the least common were burns (0.4%) and blood poisoning (1.9%). Nearly 72 percent of all accidents involved hands, wrists, and fingers. Ankles, feet, and

Table 7.9
Percentage of accidents by type and body part for well workers
(N = 46)

Body part	Type of accident										
	Bruising/ blows	Wounds	Stabs	Sprains/ strains	Fractures	Crushing	Burns	Frostbite	Embedded foreign bodies	Blood poisoning	Total
Eyes	–	–	–	–	–	–	–	–	–	1.9	1.9
Head and neck	1.2	–	–	–	–	–	–	–	–	–	1.2
Back	–	–	–	1.5	–	–	–	–	–	–	1.5
Chest	.4	–	–	–	.7	–	–	–	–	–	1.1
Stomach and lower abdomen	–	–	–	–	–	–	–	–	–	–	–
Arms	.7	–	.4	.4	1.5	–	.4	–	.4	–	3.8
Hands and wrists	.7	8.5	.7	2.9	5.6	.4	–	6.7	.4	–	25.9
Fingers	.4	21.0	.7	.4	2.9	16.0	–	.4	3.7	–	45.5
Legs	–	–	–	–	1.2	1.5	–	–	.4	–	3.1
Ankles and feet	.7	–	.4	1.2	.7	6.3	–	–	–	–	9.3
Toes	.4	–	–	–	.4	3.7	–	–	–	–	4.5
More than one body part	.4[1]	.7[2]	–	.4[3]	.4	.7	–	–	–	–	2.6
Total	4.9	30.2	2.2	6.8	13.4	28.6	.4	7.1	4.9	1.9	100.4

Note: Percentages in these tables may not add to 100 because of rounding.

[1] Feet and hands.

[2] Arms and back; feet and hands.

[3] In more than one limb.

Table 7.10
Percentage of accidents by type and body part for disabled workers
(N = 59)

Body part	Type of accident										
	Bruising/ blows	Wounds	Stabs	Sprains/ strains	Fractures	Crushing	Amputations	Slipped disks	Heart attacks	Collapsed lungs	Total
Eyes	–	–	1.7	–	–	–	–	–	–	–	1.7
Head and neck	1.7	–	–	–	–	–	–	–	–	–	1.7
Back	3.4	–	–	16.9	–	–	–	10.2	–	–	30.5
Chest	–	–	–	–	1.7	–	–	–	1.7	1.7	5.1
Stomach and lower abdomen	–	–	–	–	–	–	–	–	–	–	–
Arms	5.1	–	–	1.7˙	5.1	1.7	1.7	–	–	–	15.3
Hands and wrists	3.4	3.4	–	–	–	1.7	–	–	–	–	8.5
Fingers	–	3.4	–	–	–	3.4	3.4	–	–	–	10.2
Legs	–	–	–	–	3.4	–	–	–	–	–	3.4
Ankles and feet	–	–	–	3.4	1.7	5.1	–	–	–	–	10.2
Toes	–	–	–	–	–	1.7	–	–	–	–	1.7
More than one body part	3.4	–	–	1.7	1.7	1.7	–	3.4	–	–	11.9
Total	17.0	6.8	1.7	23.7	13.6	15.3	5.1	13.6	1.7	1.7	100.2

Table 7.11
Percentage of accidents by task and worker status

Type of task	Well workers (N = 46)	Disabled workers (N = 59)
Harvesting	46	68
Processing	17	17
Other tasks	17	7
Off work	20	8

toes sustained injuries in 14 percent of the cases and limbs sustained injuries in close to 7 percent of the cases.

Table 7.10 indicates the percentage of accidents by type and body part for disabled workers. Disabled workers most often suffered from back problems – slipped disks (13.6%), sprains and strains (17%), and bruising and blows (3%). As the result of accidents, disabled fishers most frequently reported pains and strains (24%), bruising and blows (17%), crushings (15%), fractures (14%), and slipped disks (14%). Over 30 percent of all accidents involved the back. Injuries to wrist, hand and fingers made up 19 percent of the cases. Ankles, feet, and toes accounted for 12 percent and limbs for 19 percent of the cases.

Tasks on board fishing vessels can be divided into three main categories: harvesting the catch; processing the catch, including its stowage; and other tasks. Accidents and injuries can occur while performing any of these tasks or while not at work – while sleeping, eating, or resting (see Table 7.11).

Accidents associated with harvesting – preparing and stowing gear, shooting or hauling gear, and handling deck machinery – most frequently happened to deckhands and other crew members and account for 46 percent of accidents reported by the well workers sample and 68 percent of those reported by the disabled sample. Injuries from such accidents are likely to be the most serious.

Accidents associated with processing fish – including gutting and cleaning, handling, and stowage of both fresh and frozen fish – accounted for 17 percent of injuries to well workers and 17 percent to disabled fishers. Besides those injuries fish handling presents specific hazards. Scales can get lodged in eyes and cause irritations. Some poisonous fish, such as redfish or sculpin, cause swelling of fingers and hands and in a few cases blood poisoning. Fish commonly bite. Cuts to fingers and hands occur frequently. All of these cuts, bites, and abrasions can become infected, given the working conditions in the fishrooms. All fishers have at one time or another had such difficulties, which they consider part of the job.

Other tasks – including keeping watch, preparing and stowing fishing gear, workshop duties, mooring the vessel, galley and pantry duties, repair and maintenance work, work in the engine-room and wheel-house, and fire and lifeboat drills – accounted for 17 percent of accidents to well workers and 7 percent of those to disabled fishers. Twenty percent of accidents to well workers and 8 percent of those to disabled workers occurred to fishers who were off duty or idle – on the dock, boarding or disembarking the vessel, off duty on the vessel, moving through the vessel, getting in and out of bed, or sleeping.

THE DISABLED FISHER

Many of the disabled men we interviewed were substantially older and in poorer health than their counterparts who were still fishing. Those disabled workers said they had known that they weren't able to keep up on the job. But they had also known that they had meagre, if not nonexistent, prospects of similarly well-paying work. Prior to their injuries, many had felt that they would inevitably suffer an accident. They recognized the relationship between their failing health and the possibility of injury but did nothing about it.

The men who leave because of work-related injuries often face long-term disability and many will never return to the fishery:

I had it in my mind all along that I was going back fishing and then the compensation cut me down ... I couldn't live on that. There's no way I could do it. So then they told me to put in for the Canada Pension. So I put in for Canada Pension and thank God I got that ... I wanted to go back to fishing. I had it in my mind I wanted to go back but my report came back from Halifax 100 percent disabled from any fishing or heavy lifting. So it hit me hard. It changed my life all over. I didn't know what to do with myself. I even thought about trying to make away with myself. (Leavers Interviews, Deckhand)

The disabled fisher faces a different future than does the healthy worker who leaves the fishery. These men commonly react to their situation with anger, longing, and despair. They did not decide to leave voluntarily, and did not prepare for drastic changes to their lives.

Sometimes long periods of physical rehabilitation may be necessary, and small fishing communities have few resources available locally. Rehabilitation centres located in Halifax or Sydney necessitate long drives or long absences from home in order to receive treatment in unfamiliar surroundings. Many of the men we interviewed spoke of having difficulty coping with life on land in general and with life in the city in particular. They endure separation from both their families – their "family" at sea and their family on land.

While they quickly said that their buddies who still fished supported them, the disabled fishers realized that they had little in common, except reminiscences about past experiences at sea. They also recognized that their families had other commitments as well. As one man said, "I never realised what the little woman did all the time I was away. Jesus, she works hard" (Leavers Interviews, Deckhand). It takes a crisis for most fishers to recognize how hard their wives work, although throughout their working lives their wives reorganize their own work schedules in order to respond to their husbands' needs and demands.

Besides physical rehabilitation, these men must attempt to put their lives back together through new employment. This usually means going back to school to complete grade 12 in order to qualify for specific retraining programs. Many of these men have not been in school for years, and when asked if they would return to school they said they "had no head for it." Most have only a grade 8 or 9 education, and some are functionally illiterate. Their basic skills relate to physical labour at sea, and they lack training in other job skills.

I heard only a handful of success stories. One fisher, confined to a wheelchair after an accident, retrained as a dental assistant, but he already had grade 12 and had been fishing for only a few years. A second man, missing a leg, did computer assembly work from his home. Another worked in a twine barn making nets. A fourth man was working toward having a gun repair business. One illiterate man could now read, and wrote poetry which he hoped to publish. But these men rarely succeed. Many turn to alcohol and drugs. Divorce occurs frequently, and we heard of two cases of injured fishers committing suicide. Many lead broken lives, plagued with physical complaints. They collect their pensions and dream of the sea, while life goes on around them.

CHAPTER EIGHT

The Stress and Strain of Fishing

DOES DEEP SEA FISHING LEAD TO A HEALTHY LIFE?

This chapter examines fishers' general health. In the previous chapter, I documented the large number of accidents and injuries for Nova Scotia deep sea fishers, especially among the crew who work in the most dangerous areas of the vessels – on the deck and in the fishrooms. Accidents do not occur randomly. The general health of workers alters the way they respond to dangerous situations. Stress and fatigue increase accident rates; healthy workers appear to be more capable of dealing with stress and more likely to avoid injury than their unhealthy counterparts. If fishers have more control over the work process and greater satisfaction with a variety of aspects of fishing, their stress levels and accident rates will decrease.

FISHERS' HEALTH PRACTICES

In our 1986 survey we asked fishers a number of questions about their general health and health practices. Fishers consider their work healthy, and most of the fishers (78%) in our sample considered themselves to be in excellent or good physical condition. Some men (16%) even thought their health had improved since they started fishing. When asked what aspects of their job promoted good health, they cited working out of doors, fresh air, and physical activity as the major reasons for their physical fitness. But about the same number (18%) thought that their health had declined once they started fishing. These men spoke of the harsh and demanding working conditions, which they found both physically and mentally stressful. They complained about chronic back problems, mental and physical stress, and constant fatigue. They also mentioned the level of physical risk · associated with their job.

I knew from ethnographic research that fishers were extremely reluctant to miss a voyage or watch on board. Besides the financial imperative, fishers said they did not want to be viewed as "an old lady," a slacker, or someone who lets the side down by not doing their fair share. In our 1986 survey, therefore, I asked fishers if they had ever missed a voyage or watch due to ill health and if they had, I asked them why. Despite the "cultural" prejudice against taking time off, 30 percent admitted to having missed a watch. Over 90 percent of the fishers sueveyed, predominately crew members, reported absences due to acute physical conditions that arose on board. Those conditions included concussions, fractured bones, severe sprains and strains, and acute gastric distress. Of the remaining cases, 7 percent reported absences due to recurring chronic conditions, usually older crew members' and officers' back problems. Three percent of the reported absences can be attributed to acute emotional problems.

Fishers appear equally reluctant to seek medical advice or treatment for minor ailments or complaints, yet 91 percent of our sample had seen a doctor in the past year! Some of the fishing companies demand an annual physical examination as a condition of employment and 19 percent of our sample had had a medical check-up for this reason. But most – 56 percent – went to have an acute work-related condition treated. Those conditions included respiratory and gastric complaints, infections, cuts, fractured bones, and muscle sprains and strains. Another 21 percent went for treatment of chronic ailments such as arthritis, or conditions associated with past injuries to backs, knees, ankles, and hands. The remaining 4 percent of the cases consisted of men who sought help for emotional or stress-related problems.

GENERAL HEALTH AND WORKING CONDITIONS

Fishers' work includes long hours at sea in a dangerous and uncontrol-lable environment, on a continually moving vessel, with constant engine noise and noxious fumes. On an open deck, usually awash with the cold water of the North Atlantic and always exposed to the ele-ments, crew members often toil in wet clothing. They also work in damp and slippery fishrooms and refrigerated holds. Cramped living and working quarters have only harsh fluorescent light. Twenty percent of the men we interviewed complained of arthritis. Over the year pre-ceding the survey, 10 percent reported having "the grippe", and just less than 25 percent reported sinus problems. Respiratory problems, including pneumonia, pleurisy, and lung infections, occurred frequently; 9 percent of our sample had suffered from one of these conditions over the preceding year. Over 10 percent of fishers endured constant motion or sea sickness. Most suffer from it, to some extent, in heavy weather.

Older members of the crew and some of the officers more frequently reported circulatory disorders, heart problems, high blood pressure, and overweight. Three percent of our sample stated that they had been treated over the preceding year for heart problems and just under 15 percent had been treated for high blood pressure.

The demanding work schedule of ten to fourteen days at sea with only two days on shore between trips continues throughout the year with only two or three breaks. This pattern combined with a watch system of either six hours on and six off, or eight hours on and four off, compounded by broken watches due to breakdowns or heavy catches, leads to chronic fatigue. Although most fishers find the work repetitive, simple, and monotonous, they must pay close attention to their actions because of the high risk involved in working with machinery on board. This high level of concentration enhances fatigue. Captains and other officers have additional pressures associated with the responsibility of running the vessel and finding and harvesting the fish. Captains often cited those pressures as reasons for leaving the fishery. Moreover, officers are more likely to take off trips or watches or to leave the fishery due to emotional problems.

The long time away from family and friends also takes an emotional toll. Fishers frequently complained about the lack of communication with their family on shore, but they seldom talked about their fears or worries. A fifth of the men we interviewed said that sometime during their career they had had a serious situation at home that had distracted them from their work at sea – a physically dangerous situation in itself. When we broke down those responses by job status, as shown in table 8.1, only 9 percent of captains and mates, compared to 19 percent of other officers and 27 percent of crew members, had ever experienced this type of problem.[1] But captains and mates talked more willingly about their worries. Only four crew members would discuss the topic, and only two of them mentioned the cause of their distress. Situations that caused stress fell into three categories: social relations, health concerns, and money problems. About two-thirds of the men mentioned problems in the area of social relations, the most common being spousal fidelity and the safety or well-being of their wives and families. Another 30 percent of the men reported health concerns, most notably acute illnesses, chronic health problems, and pregnancy of family members as their source of distress. The remaining 6 percent had been worried about money.

Table 8.1
Worry about home by job status

	Job status							
	Captains and mates		Other officers		Crew		Total	
Worry	(N)	(%)	(N)	(%)	(N)	(%)	(N)	(%)
Yes	9	9.1	13	19.4	45	26.9	67	20.1
No	90	90.9	54	80.6	122	73.1	266	79.9
Total	99		67		167		333	

Notes: Missing observations = 1. Chi–square value of 12.36 with 2 degrees of freedom, significant at $p = .002$.

HEALTH RISK BEHAVIOURS

According to Rosenbaum and Bursten (1988, 8) smoking, excessive alcohol consumption, and sedentary behaviour increase the risk of ill health, especially cardiovascular disease and cancer. Conway and co-workers (1981, 155) argued "that stress produces behaviours that are positively reinforcing in the short run but increase the long-term risk of illness. Examples of such behaviours include cigarette smoking, coffee drinking, drug consumption and overeating." In their longitudinal study of eighty US Navy company commanders, they found that coffee drinking and cigarette smoking correlated positively with perceived stress and that under high stress alcohol consumption declined but cigarette smoking and coffee drinking increased. All of these factors influence the health of fishers. In this section I will look at fishers' smoking, alcohol and caffeine consumption, and eating habits.

Sixty-seven of the sample smoked either cigarettes, cigars or pipes – 78 percent of the crew compared with 72 percent of the officers and 54 percent of the captains and mates.[2] When compared with the national (31%) or the Nova Scotia (33%) averages for males, fishermen's preference for smoking appears markedly higher (Canada, Health and Welfare Canada 1986, 5). In Canada, with its distinct regional differences, smoking rates appear highest in Quebec (37%) and the Atlantic provinces. Nova Scotia has the third highest rate of smoking for males in Canada (Canada, Health and Welfare Canada 1986, 26).

Male blue-collar workers have higher smoking rates than male white-collar workers. According to Health and Welfare Canada (1986, 6) "the highest smoking rates were found among males in transportation (42%), mining (40%), construction (39%), other crafts (39%) and fabricating (38%)." They reported smoking rates lowest in the professional (21%) and managerial occupational groups (24%), followed by outdoor occupational groups (28%) (ibid.).

Given the characteristics of their jobs, captains' and mates' work can be compared with that of male white-collar workers, and the other officers' and crew members' work can be compared with that of male blue-collar workers. However, the smoking rates of offshore fishers, whether captains, mates, or crew, remain substantially higher than those of any occupational group included in the 1986 Health and Welfare study. Yet Neutal (1989, 378–9) reported that fishers in Atlantic Canada had very low mortality rates for respiratory diseases. The rate of lung cancer was the lowest of all cancers. Given the smoking rates reported here I find this result very surprising.[3]

Kline, Robbins, and Thomas's study of captains of Gulf of Mexico shrimpers argued that smoking, over the short term, may enhance performance and reduce stress. They listed, with detailed references, a number of properties of smoking that enhance job performance and effort. Those properties included

1 enhances pleasure and relaxation (e.g., it provides a time-out and filler; is a distracter; has a palatable taste ...)
2 enhances task performance (e.g., it increases efficiency in reaction time and accuracy of signal detection tasks through sustained vigilance and alertness; improves concentration and selective attention; facilitates rapid audio and visual information processing; aids learning, memory and problem solving ...);
3 facilitates social relations (e.g., it communicates social roles and image; reinforces positive changes; aids composure ...);
4 reduces stress and negative effect (e.g., it reduces the disruptive effects of anxiety, anger, irritability, boredom and tension ...); and
5 reduces somatic problems (e.g., pain, hunger, muscular tension, and it increases energy when fatigued). (Kline, Robbins and Thomas 1989, 351–2)

Although Kline and co-workers did not analyse the smoking rates of crews of shrimpers, they did state that "work performance in the shrimping occupation is facilitated by tobacco use" (Kline, Robbins, and Thomas 1989, 352). In my study the high rate of smoking among crew members supports their hypothesis. In my ethnographic research, I found that fishers appeared to use tobacco to enhance task performance, to reduce somatic problems related to fatigue and sea-sickness, and to reduce stress.

Kline and co-workers (1989, 352) argued that offshore captains had greater levels of stress than inshore captains because they remained at sea more and away from home longer, worked at night more frequently, and had less control over their work schedules and routines, more hazardous conditions, less seaworthy boats, and fierce competition. They argued that the different rates of smoking between offshore captains (56%) and inshore captains (44%) reflected the higher level of stress. I found the frequency of smoking for Nova Scotia deep sea captains (54%) similar to that of offshore shrimp captains (56%) (Kline, Robbins, and Thomas 1989, 352). Deep sea fishers share many of the work characteristics of the offshore shrimpers and appear to share a similar adaptation to stress.

Another measure of level of stress is how much smokers smoke. Assuming that a pack of cigarettes consists of twenty-five cigarettes,[4] a national smoking survey conducted in 1985 (Canada, Health and Welfare Canada, 1986, 7, 28), reported that 16 percent of male smokers smoked one to ten cigarettes, 69 percent smoked eleven to twenty-five cigarettes, and 15 percent smoked more than a pack a day. The same study reported slightly higher averages for Atlantic Canadian male smokers; 17 percent smoked one to ten cigarettes, 65 percent smoked between eleven and twenty-five cigarettes, and 18 percent smoked more than a pack daily (ibid, 7, 29). Of those fishers who smoked cigarettes, 3 percent smoked ten or fewer cigarettes, 53 percent smoked between eleven and twenty-five cigarettes and 44 percent smoked one pack or more each day. In our earlier discussion we noted that fewer captains and mates than crew smoked. But captains and mates who smoked cigarettes smoked about the same number of cigarettes per day (31) as other officers (30) and crew members (31). Assuming that tobacco use mitigates against stress, these findings support the hypothesis that "stress produces behaviours that are positively reinforcing in the short run but increase the long-term risk of illness" (Conway et al. 1981, 155).

Besides the short-term adaptive advantage of smoking for fishers, other cultural practices encourage it. With the exception of sleeping quarters, the engine-room and around the fuel tanks, the company allows smoking on the vessel and seldom discourages it. Moreover, the nonmonetary benefits in every union contract include the right to bonded (tax-free) cigarettes. Depending on the collective agreement each man gets one or two packs of cigarettes per day at sea. Given that about a quarter of the men do not smoke, those who smoke can easily obtain more cigarettes than their ration. Thus the economic disincentive to smoke has been removed.

Drinking to relieve tension or stress appears common in many communities (Neff and Husaini 1982). Hingson and co-workers (1981) found that labourers, operatives, and skilled workers drank heavily and that boredom and stress on the job contributed to drinking. Janes and

Ames (1989) reported that in their study of blue-collar assembly-line workers the subculture of the workplace encouraged heavy drinking, which enhanced socialization among co-workers and helped workers to cope with the frustration and boredom of their job. They also found that workers with other social alternatives, particularly family and friends, did not become involved in this subculture. Cooper, Russell, and Frone (1990) argued that individuals under stress who had few personal or social resources seemed more likely to develop problems associated with alcohol use. Trice and Sonnestuhl (1988, 331) reported that Plant found that "high risk" occupations attracted heavy drinkers and (ibid., 331–2) that Whitehead and Simpkins included seafaring occupations among those occupations with a "high risk of alcoholism."

For most fishers consumption of alcoholic beverages on shore played an important role in socialization and the relief of stress[5] – 81 percent of our sample drank. Social drinking takes place in crew members' homes, lounges, taverns, restaurants, and other social forums. Some single crew members spend their whole leave "drinking with the boys." The participation of married men in these heavy drinking sessions depends on their integration into the community. Those fishers who integrate totally into family and community life seldom go drinking. For most fishers' wives these "tamed" men represent the ideal.

Few fishers reported drinking while at sea (7%). Since most companies will suspend or fire fishers who drink at sea, drinking on board generally remains a clandestine activity. In a few cases where the owners allow bonded spirits, they usually restrict the consumption of alcohol to off-watch officers and/or to crew members on the voyage home.

Rosenbaum and Bursten's (1988, 10) study based on Canada's Health Promotion data found that workers in transportation and communication sectors (11%) and in the primary industry sectors more often reported high-risk drinking behaviour – consuming four or more drinks on three or four occasions over a one-week period. Because of the nature of the work schedule – ten days on and two days off – and the prohibition of drinking at sea, it would be difficult, using this definition, to classify any fisher as practising high-risk drinking.[6] Nevertheless, we asked fishers how often they drank alcoholic beverages. Seven percent of fishers reported that they drank every day, 31 percent said they drank at least once a week, another 31 percent said they drank one or more times a month, 12 percent said they drank less than once a month and 19 percent abstained. Of those who drank every day, officers including captains and mates (6 men) were proportionally more likely to report high-risk drinking behaviour than crew members (10 men). These men compounded their risks since all of them smoked.

The nature of offshore fishers' work does not allow them to participate in organized sports or formal exercise programs. Unlike Scandinavian freighters or progressive oil rigs the vessels do not have weight or exercise rooms on board, although some men do bring small dumb-bells with them. Only 46 percent of the men said they participated in physical activities such as hockey, curling, golf, bowling, football, base-ball, badminton, tennis, karate, or dancing when ashore. This means that 54 percent did not participate in any physical activities at all when they came ashore.

I also measured sedentary behaviour indirectly by looking at what proportion of fishers weighed more than the norm. We used the Ponderal Index, defined as $((weight)/(height)^2)100$, where weight is in kilos and height is in centimetres, to assess what proportion of our sample appeared under, over, or at normal weight. Ponderal Index values between .246 and .295 lie within the norm, with values below .246 tending toward underweight and values above .295 tending toward overweight (Bray 1982, 24). This index gives only rough approximations but it does allow us to look at trends within the population under study.

Table 8.2 shows the distribution of the Ponderal Index by job status.[7] Although about one third of all fishers weighed within the norm, 45 percent of officers other than captains and mates and over 50 percent of captains and mates appeared to be overweight. About one-third of the crew seemed to be overweight, and another third seemed to be under-weight. The sedentary nature of wheel-house work, the long tedious hours of watch, and the emotional stress of being responsible for the vessel, crew, and catch all contribute to overweight captains, mates, and other officers. But sometimes mates and captains must perform physically demanding tasks. When this happens those men who are overweight and/or out of condition find it difficult to perform appropriately. Some captains and mates recognize the situation and delegate the task to an appropriate crew member or officer; however, many attempt the task and put both themselves and others in danger.

This measure also looks indirectly at eating habits. Each man has particular likes and dislikes, and trying to satisfy the majority proves a difficult chore, especially if the cook has no training in food preparation or nutrition. A few, ex-navy cooks or former stewards, have had some training, but most cooks have learned their trade on board. Beef, pork, bacon, sausages – anything but fish – constitute the main courses. Few meals feature fresh vegetables and fruits. Many of the

Table 8.2
Ponderal Index by job status

Ponderal Index	Job status							
	Captains and mates		Other officers		Crew		Total	
	(N)	(%)	(N)	(%)	(N)	(%)	(N)	(%)
Underweight (below .246)	9	9.1	17	25.8	54	32.1	80	24.0
Normal weight (.246–.295)	38	38.4	19	28.8	58	34.5	115	34.5
Overweight (above .295)	52	52.5	30	45.5	56	33.3	138	41.4
Totals	99		66		168		333	

Notes: Missing observation = 1. Chi-square value of 20.79 with 4 degrees of freedom, significant at $p = .00035$.

meals include deep-fried, greasy, and starchy foods – partly because they are relatively easy to prepare and partly because most men like them. With heartburn a prevalent complaint, fishers comsume antacid tablets like Rolaids by the jar. As one man (Well Workers Interviews, Deckhand) explained, "We go through those great big bottles of Rolaids. Eighteen fellows goes through six of them in a trip. So what's in a bottle? I'm not right sure but that's a lot of Rolaids to go through on a trip."

Fishers often complain of gastric problems such as upset stomachs, heartburn, wind, and gas. Over 28 percent have stomach problems. Neutal (1989, 378) reported in her study on commercial fishers of Atlantic Canada that those workers had higher than expected mortality from stomach cancer. She cited a diet high in fat and fried food, the consumption of large quantities of caffeinated beverages and alcohol, and motion sickness as contributing to this high mortality rate. Cooking for a crew of sixteen men presents its own difficulties. About 1½ percent of our sample suffered from diabetes and required a special diet. Those men found it particularly difficult to eat properly.

In our study we asked fishers how many cups of caffeinated beverages – coffee, tea, cola – they drank each day. Only 10 percent of fishers did not drink caffeinated beverages. On average captains/mates drank nine cups a day, while other officers drank an average six cups and crew members an average seven cups a day.[8] Caffeine, a diuretic, stimulates the heart and nervous system and may be associated with heart palpitations (Wilson et al. 1991, 1:14). Caffeine also lowers blood pressure, relieves muscle tension, abolishes fatigue, and maintains wakefulness. As few as three to five cups of coffee can significantly disturb sleep patterns and higher levels of consumption can cause insomnia (Wilson et al. 1991, 1:213). The consumption of ten cups of coffee or its equivalent a day can cause behavioural effects (Wilson et al. 1991, 1:424). My ethnographical research indicates that captains and mates and, to a lesser extent, other crew members use caffeinated beverages to combat fatigue and drowsiness. Neutal (1989, 378) reported that cancer of the pancreas had the highest mortality rate of all cancers occurring in North Atlantic fishers. She cited consumption of alcohol and caffeinated beverages and smoking as risk factors.

STRESS AND WORKERS' PERCEPTION OF RISKS

In the preceding chapters on job satisfaction and working conditions I identified a number of negative aspects of fishing – the physical environment, working conditions, work organization, and work schedule. Riordan, Johnson, and Thomas (1991, 391) found that safety concerns, long watches at sea, and having responsibility for others' lives contributed significantly to stress among fishers. Kline, Robbins, and Thomas (1989, 352) defined "Fishermen's Stress Syndrome,"[9] a combination of eight major workload stressors that directly or indirectly affected fishers' health. The stressors they identified consist of

1 overload, or excessive quantity and quality of workload;
2 underload, or inadequate mental stimulation;
3 shift work;
4 migration or frequent alternation between land and sea;
5 social interaction stressors (e.g., overcrowding);
6 working conditions at sea;
7 organizational role; and
8 career development.

All of these factors affecting fishers' general health resemble the items I identified as factors influencing job satisfaction.

I assume that fishers' level of stress depends on their perception of the daily risks they run at sea. Other studies indicate that the definition of risky behaviour can be influenced by social and cultural values (e.g.,

Douglas 1986). If risk perception proves to be a social as well as an individual phenomenon, then we must analyse the social factors that contribute to fishers' awareness of physical risks and identify which experiences and training techniques increase this awareness. Fishers consistently deny the dangers inherent in their work. That type of short-sightedness and inattention to risks appears to be not so much a sign of weakness as an attempt to protect certain values and their concomitant social institutions. Fishers' denial of risks may be a way to relieve stress. Conversely, awareness of those risks may be mal-adaptive in some way.

Fishers belong to a subculture that glorifies danger. Tales immortal-ized in song, art, and stories celebrate these fearless men confronting the elements. Being part of a culture that honours courage and daring influences fishers' perceptions of risks and permeates the fishing subculture. For example, during our study an inshore fisher hired two young men through Canada Manpower to go fishing with him. Neither of these fellows had ever gone fishing before, nor did they come from fishing families. While fishing a few miles from shore using long lines, one of the young men had an accident: a six-inch hook went through the palm of his hand. When the captain went to remove the hook, the young man drew his fishing knife on the captain and forced him to return to port. The captain charged the man with mutiny. When the community discussed the case, support for the captain and for the young man split down lines of "occupational community." Those people involved in the fishery or from fishing families sided with the captain. They argued that such an injury occurred frequently at sea. If the captain had been allowed to remove the hook, then the voyage could have continued, and if the wound remained troublesome at the end of the trip the young man could have gone to the hospital. They thought it inconceivable for the vessel to return for such a minor problem. People from nonfishing backgrounds or households sympa-thized with the young man. Surely a six-inch hook through the palm of your hand required immediate medical attention. The case was eventually thrown out of court on a technicality, but it does illustrate how people from this fishing culture perceive risks at sea.

Displaying a "macho" attitude remains prevalent throughout the fishery. Fishers routinely dismiss risks at sea as minor and trivialize the injuries and accidents they sustain during their work. When I pretested the 1986 survey I included the following question, "Have you ever had a serious accident?" All the workers pretested answered "no," yet the interviewer noted that all had sustained at least one injury that could be deemed serious. One man explained, "You know you stop fishing after you have a serious accident." Since those men could still fish, they felt that they had never had a serious accident. I modified the question to read "Have you ever had an accident on board a fishing vessel?" That solved our problem with consistency. I also added an additional question – "What are routine accidents?" – and their answers ranged from infections and blood poisoning (caused by cuts from fish scales and ocean perch spines) to sprains and fractured limbs from slips and falls.

Whenever we asked about physical risks, fishers attempted to make the extraordinary ordinary. Once the situation had been limited to the ordinary their anxiety decreased, and they felt capable of controlling the situation.[10] Although fishers will seldom admit that they themselves take risks, they will admit that other fishers do. When asked why those men get into dangerous situations, the most frequent answers focused on carelessness and inattention to their work. The perception of being able to control situations at sea allows fishers to accept the risks and to differentiate themselves from injured workers.

Fatalism also underlies fishers' perception of risks.[11] For example, very few fishers can swim. When asked why they do not learn, they typically answered, "Why learn to swim when it would just take you longer to drown?" Before the modern survival suit, no one could survive for more than a few minutes in the cold waters of the North Atlantic. Fishers believed that if they fell overboard, there would be only a few situations in which a worker could not be fished out immediately (e.g., bad weather, very high seas, sinking vessel).[12] Most fishers knew that if a boat sank the best chance of being rescued lay in holding onto debris from the vessel: if you could swim, then you might try to reach shore or do something equally stupid. Many felt that if the weather deteriorated enough to sink the boat there would be no other boat close enough to save them, because rescue helicopters/planes would not be flying, and the water was too cold to survive in whether one could swim or not. Those expressions of fatalism reflect the perception of being unable to control the forces of the sea and an acceptance of the things they cannot change. Thus the denigration of risks and a fatalistic attitude can be interpreted as complementary protective mechanisms, which make working on board acceptable by allowing the fisher to be both physically and psychologically prepared for work in this dangerous environment.

Sometimes the high levels of stress associated with the dangerous working environment overwhelm psychological mechanisms for coping. Three common manifestations of mental stress at sea consist of "the channels," "going on a tear," and "breaking up." Men speak of "having the channels" when they sail for home. They feel restless, impatient to get ashore, and cannot sleep. Poggie (1980, 128) reported Gloucester fishers referring to "channel fever," which has similar symptoms. Many of the men we spoke to who had had "the channels" attributed them to being too long at sea and needing a break.

"Going on a tear" usually appears after an extremely stressful trip, such as one with a major fire or an accident to a shipmate on board. But it can also occur after a series of trips with moderate levels of stress. The symptoms resemble "the channels," but in this case the fisher finds it difficult or impossible to adjust to shore life. He stays awake; he cannot settle down or sleep. He may get in his car and drive for miles going nowhere in particular. He may go to a bar or Legion hall and drink until he passes out. He talks about trying to catch up, and being all "broke-up." For some men, usually single or separated, this can become a way of life. For most, "being on a tear" remains an occasional occurrence. Nevertheless, some fishers' wives find greeting their husbands at the dock a stressful experience. As one wife said, "I never know what it is going to be like when he comes home. Mostly he's good ... But sometimes it's like an animal let out of a cage ... When he's like that I take the kids and go to Mom's" (Wives Interviews). In situations where this kind of behaviour continues, most women find it impossible to cope with the stress and end up leaving the marriage.

In interviews, fishers would describe men "breaking up" when they could not cope with the pressures at sea. "Breaking up" took many forms. In extreme cases fishers suffered mental breakdowns. One man calmly packed his duffle, told the captain he was going home to speak to his wife but would be back, and then walked off the stern carrying his bags. More commonly, "broken" men simply become "nervous." Some men appeared edgy about machinery, others worried about the boat sinking and would wear their survival suits all the time, still others seemed anxious while shooting away or hauling in the drags.

These "broken men" recognized the dangers of fishing, but this awareness had caused them to overreact to those risks. They no longer felt in control of the situation and could not deal with their working environment. In some cases, their nervousness resulted in a serious accident that ended their fishing career. In other cases the fisher consciously recognized his fear and dropped out.

Such anxious behaviour makes the average fisher uncomfortable because it reminds him of the dangers at sea, and may undermine his perception of his ability to control his working environment. Ideas of masculinity, particularly the feeling that real men can "take it" (whatever "it" is), militate against those men asking for or getting any support and help from other crew members.

The term "broken up" not only refers to unusual behaviour but also explains why accidents happen. When talking about injured men, fishers would often say a man was "broken up," implying this as a reason for his accident. Most fishers thought "broken" men dangerous because they could not be trusted to perform predictably. Some even thought "broken" men so dangerous to themselves and others that they should be barred from sailing.

Once men leave the fishery their perceptions of the industry's physical risks change, especially if their reason for leaving involved an injury. Although leavers listed a range of routine injuries similar to those listed by current fishers, they talked about the serious nature of those injuries, and they would discuss more openly risks at sea that they had trivialized or repressed while working on board. Although some realized how little control they had had at sea, they could not talk to their ex-shipmates about it. As one man (Leavers Interviews, Deckhand) said, "I can talk to you about this. My buddies, they won't listen, they can't listen. If they do, they'll break up." Another disabled man said he could not talk about his accident with old shipmates because it made them nervous and they would stop coming to visit him. These ex-fishers respect and realize why fishers trivialize risks, and they discuss those risks because they will never go fishing again.

CHAPTER NINE

Safety Awareness

In order to change fishers' approach to safety and to develop effective educational and training programs, those involved must understand fishers' current attitudes to safety issues and how they fit into the fishers' subculture. In the preceding chapters I have argued that fishers' working conditions, attitudes to their work, experience, family situations, physical and mental health, and perceptions of risks and danger on the job all influence their level of safety. This chapter examines fishers' current levels of satisfaction with safety, describes what factors underlie those attitudes, and discusses how awareness of safety issues can be enhanced.

LEVELS OF SATISFACTION WITH SAFETY

How satisfied are fishers with safety? For this analysis, "the level of satisfaction with safety," the dependent variable, had a five-point scale. The categories originally ranged from "very dissatisfied" (1), "dissatisfied (2)," "neither satisfied nor dissatisfied (3)," "satisfied (4)," and "very satisfied (5)." Because of the small number of "dissatisfied" and "neutral" cases, for later analysis I collapsed the levels of satisfaction into three categories: "satisfied," "neutral," and "dissatisfied."

On the whole, Nova Scotia deep sea fishers appear very satisfied with the level of safety in their work. As indicated in table 9.1, just over 85 percent of workers seemed "satisfied" or "very satisfied," while less than 9 percent appeared "dissatisfied" or "very dissatisfied" with the current level of safety. Pollnac and Poggie's study (1990, 408) found that 60 percent of New England fishers reported being "satisfied" or "very satisfied" with the level of safety, while only 20 percent appeared "dissatisfied" or "very dissatisfied."

Table 9.1
Distribution of levels of satisfaction with safety

Level of satisfaction with safety	Frequency	Percentage
Very dissatisfied	7	2.1
Dissatisfied	22	6.6
Neutral	18	5.4
Satisfied	154	46.2
Very satisfied	132	39.6
Total	333	99.9

Note: missing values for 1 case.

Table 9.2
Analysis of variance across levels of satisfaction with safety categories

Safety variable	Levels of satisfaction with safety			F Ratio	DF	p
	Dissatisfaction	Neutral	Satisfaction			
Age[1] (years)	36.9	34.9	39.8	2.7	2	.07
Dependents[2]	2.7	2.3	3.0	2.7	2	.07
Education[3]	8.7	9.4	8.4	2.4	2	.09
Experience[4]	15.1	12.9	17.4	2.4	2	.09

Note: As measured by analysis of variance (F-test) with probability levels indicated in the following manner: $*p \leq .05, **p \leq .01, ***p \leq .001$.

[1] Missing values for 2 cases.
[2] Missing values for 1 case.
[3] Missing values for 4 cases.
[4] Missing values for 2 cases.

Although Nova Scotia fishers expressed a high general level of satisfaction with safety, it was necessary to find out if it varied relative to demographic characteristics.[1] I conducted an analysis of variance of demographic variables – "age," "number of dependents," "education level," and "years of experience" – with three categories for "satisfaction": – "satisfied," "neutral," and "dissatisfied." All of the demographic variables derived from direct questions. As indicated in table 9.2, differences between the categories did not indicate statistical significance – a finding similar to Pollnac and Poggie's results (1990, 409).[2]

Younger and less experienced men tended to be more critical of safety on board vessels. Poggie and Pollnac (1990, 409) suggested that this depended on three factors: first, the most dissatisfied workers left the occupation after only a few years of work; second, those with greater experience had developed more effective coping mechanisms; third, workers had become habituated to the conditions. To those factors, we would add three more specific to the Nova Scotia deep sea fishery.[3] Younger and less experienced workers, usually deckhands, do the most dangerous work and take the greatest risks (this helps to explain lower levels of satisfation with safety reported by crew). These workers seek to enhance their economic position by taking safety courses, including first aid, emergency safety, and firefighting. Those courses did not exist when older workers started fishing, and by the time they take a "refresher course," they have become desensitized to the issues. Finally, companies have increased their attention to safety and markedly improved conditions on many vessels. Old hands recognize those improvements but younger workers do not.

But the acceptance of the current levels of safety reflects more than the lack of awareness of dangers. The possibility of accidents and the resulting injuries can be seen as part of the price paid for being a fisher, and the trivialization of risks can be seen as an unconscious decision to trade off safety concerns for economic ones. For most fishers, economic uncertainty – price of fish, quantity and quality of fish caught – poses a much greater threat to survival. With the depletion of the stocks, those risks take on greater importance. Thus monetary benefits (e.g., steady employment, higher wages) appear to be adequate compensation for the lack of nonmonetary benefits (e.g., safety, independence). If, as Smith (1988) argues, economic concerns prevail over safety concerns, then workers will take dangerous employment with high wages in order to meet their immediate economic needs. If they accept this trade-off, they must be satisfied with the current level of safety.

AWARENESS OF FACTORS
AFFECTING SAFETY

By surveying the literature on physical risks in the North Atlantic fisheries and their association with technology, technological change, and working conditions, I identified sixteen items that influence safety. In both the 1986 and 1987 surveys I asked fishers to rank their assessment of the importance of those sixteen items to safety on board their vessels. I measured the level of importance of various working conditions and attitudes about safety by using a five-point scale – "very important (1)," "important (2)," "neither important nor unimportant (3)," "unimportant (4)," and "very unimportant (5)." Because of the small number of cases in the "very important" and "very unimportant"

Table 9.3
Rank order of distribution of levels of importance of safety variables

Safety variable	Levels of importance of safety						Mean	Missing
	Important		Neutral		Unimportant			
	(N)	(%)	(N)	(%)	(N)	(%)		
Machine operator	333	99.7	1	0.3	0	0.0	2.00	0
Individual's attitude	329	98.8	3	0.9	1	0.3	2.02	1
Maintenance of machinery	329	98.5	3	0.9	2	0.6	2.02	0
Fellow workers	326	97.6	6	1.8	2	0.6	2.00	0
Mental pressure	321	96.4	6	1.8	6	1.8	2.05	1
Captain's attitude	319	95.5	11	3.3	4	1.2	2.06	2
Fatigue	311	93.7	11	3.3	10	3.0	2.09	2
Location of machinery	292	89.3	20	6.1	15	4.6	2.15	7
Drinking	293	88.8	20	6.1	17	5.2	2.16	4
Time at sea	291	87.7	25	7.5	16	4.8	2.17	2
Luck	262	79.9	27	8.2	39	11.9	2.32	6
Government regulations	247	74.8	38	11.5	45	13.6	2.39	4
Crew size	227	68.2	31	9.3	75	22.5	2.54	1
Horseplay	215	64.8	37	11.1	80	24.1	2.59	2
Size of catch	172	52.6	56	17.1	99	30.3	2.78	7
Company's attitude	172	51.8	31	9.3	129	38.9	2.87	2

categories, for this analysis I collapsed the levels of importance into three categories: "important," "neutral," and "unimportant."

Table 9.3 presents those variables, as ranked by fishers from the 1986 sample (current workers). Over 95 percent of workers indicated six items – "the machine operator," "an individual's attitude to safety," "maintenance of machinery," "fellow workers," "mental pressure," and "the captain's attitude" – as important for safety. More than 90 percent of fishers identified "Fatigue" as important. More than 85 percent of workers recognized that "location of machinery," "drinking," and "length of time at sea" influenced safety. Eighty percent deemed "luck" to be important. Less than 75 percent of the sample distinguished "government regulations," "crew size," "horseplay," "size of catch," and "company's attitude" as contributing to safety. Over 30 percent of the workers reported "size of catch" and "company's attitude" as unimportant.

In order to examine the interaction of workers' attitudes to the importance of safety variables with the level of satisfaction with safety, I conducted an analysis of variance. The results, in table 9.4, indicate three statistically significant relationships associated with three variables – "individual's attitude," "maintenance of machinery," and

Table 9.4
Analysis of variance of attitudes to safety variables crosstabulated with levels of
satisfaction with safety

Safety variable	Levels of satisfaction with safety			F Ratio	DF	p
	Dissatisfaction	Neutral	Satisfaction			
Machine operator	2.0	2.0	2.0	.08	2	.92
Individual's attitude	2.1	2.0	2.0	6.14	2	.00***
Maintenance of machinery	2.2	2.1	2.0	12.67	2	.00***
Fellow workers	2.0	2.1	2.0	1.77	2	.17
Mental pressure	2.1	2.0	2.0	1.51	2	.22
Captain's attitude	2.1	2.1	2.0	.86	2	.43
Fatigue	2.1	2.2	2.1	1.38	2	.25
Location of machinery	2.4	2.2	2.2	5.10	2	.00***
Drinking	2.4	2.1	2.1	2.43	2	.09
Time at sea	2.1	2.1	2.2	.87	2	.42
Luck	2.3	2.7	2.3	2.54	2	.08
Government regulations	2.6	2.6	2.4	2.31	2	.10
Crew size	2.5	2.4	2.6	.45	2	.64
Horseplay	2.9	2.5	2.6	1.62	2	.20
Size of catch	2.8	2.7	2.8	.04	2	.96
Company's attitude	2.8	2.5	2.9	1.16	2	.31

Note: As measured by analysis of variance (F-test) with probability levels indicated in the following
manner: $*p \leq .05$, $**p \leq .01$, $***p \leq .001$.

"location of machinery." In all cases those fishers who reported the
lowest levels of satisfaction with safety found the safety variables more
important than those workers who reported higher levels of satisfaction with safety.

In my analysis, workers attributed greater importance to those factors directly associated with their work environment and working
conditions (e.g., machine operator, fatigue, etc.). Fishers deemed
factors that affected safety but appeared distant from the immediate
workplace less important or unimportant (e.g., government regulations, company's attitude), though they did recognize that those same
factors influenced job satisfaction. Yet many of those latter factors
objectively influence the working conditions and the level of awareness
of safety.

WHERE SAFETY CONCERNS
ORIGINATE

If in follow-up interviews fishers mentioned particular safety or health
concerns, we asked them, "How did you become aware of these problems?" The responses indicated a broader range of opinion than the

statistical analysis above would suggest. Their responses fell into five categories:

1 from other workers on the job;
2 union activities;
3 education/training in the form of courses;
4 government and company demonstrations; and
5 having had a serious accident or having seen a fellow crew member be seriously injured or killed.

Being Responsible for Each Other

Fishers most commonly learn about safety and dangers at sea from other men on the job. Although the Large Vessel Owners' Association sponsors a deckhand's course, fishers require no formal training. As one captain explains,

When a man comes aboard, we call 'em greenhorns. You have to take 'em around and show 'em. But that is just once around the deck or twice around the deck. But if that man would sit down and watch films for a week or two before he went out then he'd know a lot more of what's going on. (Well Workers Interviews, Captain)

After this quick introduction, one of the old hands will be assigned to look after the greenhorn. Learning the job and the dangers associated with it depends on the knowledge of other workers. To train those new men requires skill and knowledge. As the captain (Well Workers Interviews) quoted earlier explained, "It's not like on land. Some places you work on land, you got four men for one job; out there you got one man for twenty jobs." Greenhorns initially do simple tasks, such as sorting the fish or standing watch, and spend their "spare time" learning basic skills – tying knots, making nets, etc. As they gain skill and confidence they take on more complicated and difficult tasks. When the greenhorn has put in a minimum of three months of sea time and has his captain's approval, he can try for his deckhand papers, which will allow him full crew status.

I asked fishers what made a good crew. Invariably they would talk about taking responsibility for your fellow crew members and looking out for them. As one deckhand explained,

A good crew works and sticks together. They can protect each other, watch out for each other, and everything runs so much smoother ... Like I said, everybody watches out for everybody else. There's the odd guy that's kind of careless, hyper, stuff like that. So if he makes a little mistake he don't notice if you go behind and correct him. But you don't want to do it in front of the man's face. It's very important that everybody digs in like and does what's right. (Well Workers Interviews, Deckhand)

This feeling of commitment to each other enhances their perception of being able to control their working environment. But when the make-up of the crew changes (i.e., a new but experienced crew member, greenhorn, or acting skipper joins the ship), the crew must again develop a sense of trust before they will feel secure. As one captain stated,

It is a big worry. If we got our old crew for trip after trip, it's fine. You get one greenhorn different on one of those ships and it's a different ball game. Because it makes the system a little bit different. I'm not saying he's not a good man, that's not the point. The point is he doesn't know like everybody else. He works a little different and everybody has a different system. (Well Workers Interviews, Captain)

Predictable behaviour of crew members is essential for fishers' feeling of well-being.

Unions and Companies Promoting Safety Awareness

If companies and unions display a positive approach to safety, fishers' awareness of these issues should be enhanced. A chi-square analysis, presented in table 9.5, shows the relationship between levels of satisfaction with safety and variables associated with "union membership," "union type," and "vessel type." Two items, "vessel type" and "union membership," appear to be statistically significant.[4] Workers on trawlers reported the lowest level of satisfaction with current safety practices: over 10 percent indicated dissatisfaction, another 10 percent appeared neutral, and just under 80 percent indicated satisfaction. Workers on mixed-gear vessels reported the highest levels of satisfaction, at just under 95 percent. Scallop workers fell in the middle; almost 88 percent reported satisfaction, while just over 9 percent did not.

Those workers belonging to a union appeared more likely to be dissatisfied with the current level of safety than nonunionized fishers. A study by Suschnigg (1988) compared three integrated steel mills located in Ontario, dividing unions into two types, militant and moderate. He found that the mill workers belonging to militant unions

Table 9.5
Chi-square analysis of descriptive variables crosstabulated with levels of satisfaction
with safety

| Safety variable | Levels of satisfaction with safety | | | N | χ^2 | DF | p |
	Dissatisfied (%)	Neutral (%)	Satisfied (%)				
Union							
Yes	9.5	7.9	82.6	190	5.73	2	.05*
No	7.8	2.1	90.1	141			
Type of union							
Militant	10.8	8.1	81.1	111	.84	2	.66
Moderate	7.6	6.3	86.1	79			
Vessel type							
Scalloper	9.2	3.3	87.6	153			
Trawler	10.2	10.2	79.7	118	11.10	4	.03*
Mixed	3.4	1.7	94.9	59			

Note: As measured by chi-square test with probability levels indicated in the following manner:
*$p \leq$ .05, **$p \leq$ .01, ***$p \leq$ .001.

had fewer accidents and paid more attention to safety than those belonging to more moderate unions. For the purposes of this study, I labelled United Fishermen and Commercial Workers (UFCW) and Newfoundland Fishermen and Allied Workers (NFAW) as militant unions and all the other unions as moderate, based on self-definition by the unions involved.

I used chi-square analysis to test the hypothesis that members of the more militant fishers' unions would be less satisfied with their current level of safety because of their greater safety consciousness. Table 9.5 also presents the results of this analysis. Although the findings appear not to be statistically significant, the trend supports the hypothesis.

In order to find out what fishers thought of their company's attitude to safety, the interviewers asked them how interested their company was in safety. The categories included "interested (1)," "neither interested nor uninterested (2)," and "uninterested (3)." Next I made an analysis of the variance of companies' attitudes to safety by the level of satisfaction with safety. For that analysis, I measured "the level of satisfaction with safety," the dependent variable, on a three-point scale. The analysis showed that those fishers who felt that their company cared about safety appeared statistically more satisfied with their level of safety.[5]

In my analysis "union membership," "company's attitude to safety," and "vessel type"[6] proved to be factors that enhanced fishers' awareness of safety. Fishers' unions always display concern for safety, but wages and job security have had priority in the past. Since the early 1980s safety has taken on a higher profile in collective bargaining, corresponding to an increased awareness and interest in safety on the part of both unions and companies.

The largest company in Nova Scotia, National Sea, has hired a fleet manager with a special mandate for safety. National Sea arranges demonstrations of safety equipment and procedures, offers first aid classes to all employees, and has made monthly safety inspections and fire drills on vessels mandatory. Fishers earn tickets (licences) through a combination of sea time and classroom instruction based on written and practical materials. This company supports fishers who want to advance their training, particularly officers who wish to upgrade their tickets to Master Mariner I (Fishing). This increased attention to safety matters had several causes: a series of accidents resulting in fatalities in the early 1980s, increased Workers' Compensation Board premiums, increased union pressure for better working conditions, and the move to scientific management.

Some companies have followed this lead; many have not. However, when fishers leave this company to work for other employers they take their awareness of safety issues with them. Those fishers, working with their unions or worker groups, have helped to raise the consciousness of their fellow workers and to demand safer working conditions from their employers.[7]

With the push by unions and some companies to increase safety awareness, there has been a corresponding increase in the demand for first aid courses. In the progressive companies, all crew members take those courses, usually early in their careers. Only ambitious workers, usually after two or three years of sea experience, take advanced courses. As discussed above, younger and less experienced workers tend to be less satisfied with safety, and the interaction of those factors cannot be separated.

The Impact of Training Programs and Demonstrations

While many of the fishers interviewed minimized the risks, others appeared more aware of the dangers. Two "graduates" of their company's required safety course gave the following statements:

There's risk to your health wherever you're at. You're just as safe on the water as you are with anything else ... It's the same as on land, sometimes you get hit with a truck, where on the water you don't get hit by a truck, you get bumped on the rail. So it's a little safer there. Just watch yourself. (Well Workers Interviews, Deckhand)

Oh, when it's rough weather you can't say that there is no risk 'cause there is. You put your foot aboard the boat and you're taking a risk about the same as in the woods logging. You got that risk every day that the tree you're sawing down is going to fall on top of you. It's about the same as going fishing. You're taking a risk as soon as you step aboard the boat of going overboard or whatever. You know that there is no control over it ... If you have got a good gang that you know that's conscious of what they are doing all the time, you don't get hurt. Like I say most accidents happen out of neglect or stupidity and if everybody keeps their mind on what they're doing – no horsing around and stuff – you don't get hurt. (Well Workers Interviews, Deckhand)

These men, as did others, compared the risks at sea with those on land. Many fishers told the interviewers that they considered fishing safer than driving a car, stating that fewer people died at sea than on Nova Scotia highways, but they failed to take into account the total numbers at risk. Woods workers represent a comparable group, since those occupations have an equivalent rate of injury. Poggie (1980) reports a similar comparison of relative risk by New England fishers. When he suggested that fishing entailed physical risks, the fishers he spoke to either denied it or said that "it is no more dangerous than riding in a car" (Poggie 1980, 123) – an attitude Poggie attributes to "repressing their awareness of the dangers of their occupation."

Many workers said that they had taken courses that had a large safety awareness component. When "drilled" by the interviewer they knew all the right answers, but they did not all believe that they could be helped by this knowledge in the event of an accident. So what makes the difference? Table 9.6 presents a chi-square analysis of variables influencing safety awareness by levels of satisfaction with safety. I assessed those measures through direct questioning. Only one relationship represented a statistically significant finding – "has taken a first aid course." Workers who had taken a first aid course reported less satisfaction with safety aboard their fishing vessel. "Has private insurance," "has taken a course," or "wears safety clothing" showed no statistical difference when their level of satisfaction with safety was measured. Depending on its timing and content, taking a first aid course, training facilitated by its universality in some companies (an indication of a company's commitment to safety), can enhance a fisher's awareness of safety.

If safety courses remain voluntary, only the workers who have been sensitized to the issues will take them. Compulsory courses given by

Table 9.6
Chi-square analysis of measures of awareness of safety crosstabulated with levels of satisfaction with safety

Safety variable	Levels of satisfaction with safety			N	χ^2	DF	p
	Dissatisfied (%)	Neutral (%)	Satisfied (%)				
Has private insurance							
Yes	8.3	6.5	85.3	278	4.01	2	.13
No	10.9	–	89.1	55			
Has taken a course							
Yes	9.3	6.6	84.1	227	2.25	2	.32
No	7.7	2.9	89.4	104			
Has taken a first aid course							
Yes	10.9	8.0	81.0	174	7.63	2	.02*
No	6.3	2.5	91.1	158			
Wears safety clothing							
Yes	8.6	5.9	85.5	303			
Sometimes	8.3	–	91.7	24	1.69	4	.79
Never	–	–	100.0	1			

Note: As measured by chi-square test with probability levels indicated in the following manner: *$p \le .05$, **$p \le .01$, ***$p \le .001$.

government agencies, companies, or unions that upgrade a worker's qualifications reach a broader audience and have better potential for increasing awareness. But compulsory courses may still not change attitudes substantially; for some fishers, the denial of control seems adaptive from a psychological viewpoint. In order to tailor courses that meet both the physical and psychological needs of fishers, people must understand what function the denial of danger has for them. Fishers' anxiety can be reduced by teaching how to reduce risks without undermining the perception of being in control of the working environment.

The Impact of Exceptional Risks

Workers who have experienced exceptional risks at sea may be more conscious of safety. More than 75 percent of fishers who had had an accident said that they were more aware of safety concerns afterwards. I used chi-square analysis to compare different measures of experience of exceptional risks with levels of satisfaction regarding safety. Those measures derived from direct questions and the results appear in table 9.7. Only one case proved statistically significant – "ever been in a major fire at sea." Over 17 percent of those who had experienced

Table 9.7
Chi-square analysis of measures of exceptional risks at sea crosstabulated with levels
of satisfaction with safety

Safety variable	Levels of satisfaction with safety						
	Dissatisfied (%)	Neutral (%)	Satisfied (%)	N	χ^2	DF	p
Ever been in a major fire at sea							
Yes	17.1	5.3	77.6	76	8.68	2	.01**
No	6.3	5.5	88.3	256			
Ever abandoned vessel							
Yes	10.9	4.3	84.8	46	.39	2	.82
No	8.4	5.6	86.0	286			
Ever been ill at sea							
Yes	12.8	4.3	83.0	94	2.92	2	.23
No	7.1	5.9	87.0	239			
Ever been in an accident at sea							
Yes	8.4	5.8	85.8	155	.12	2	.94
No	9.0	5.1	86.0	178			
Ever been on an unsafe vessel							
Yes	11.1	4.9	84.0	162	2.36	2	.31
No	6.4	5.8	87.7	171			
Ever collected WCB							
Yes	9.4	4.7	85.8	106	.23	2	.89
No	8.4	5.8	85.8	226			

Note: As measured by chi-square test with probability levels indicated in the following manner:
$*p \leq .05$, $**p \leq .01$, $***p \leq .001$.

a major fire at sea reported dissatisfaction with safety, compared to just over 6 percent of those workers who had not. Levels of satisfaction with safety for the other variables – "ever abandoned vessel," "ever been ill at sea," "ever been in an accident at sea," "ever been on an unsafe vessel" and "ever collected Workers' Compensation (WCB)" – appeared not to be statistically different.

Given our experience with the "serious accident" question,[8] the team asked fishers in follow-up interviews if they considered items in the "exceptional experience" category to be extraordinary. They associated experiences such as falling overboard, abandoning a vessel, or having an accident at sea with dangers regularly encountered during fishing. Fishers consider a fire at sea or witnessing the death of a co-worker an extraordinary event. Those exceptional experiences critically affect safety awareness because they make workers rethink their basic assumptions about safety at sea.

Wives' Worries and Concerns

While fishers trivialize the risks they take, their wives characterize their lives as being full of uncertainty and fear that their husbands may return maimed, or not return at all. This common concern binds crew members' wives together. As one fisher's wife described it,

They burn up the phone lines ... I can't stand the phone ringing like that, and hearing the same thing. It'll be the wives calling. The company couldn't stand it either, they just about drove management crazy ... They had to set up a number that you can call in the mornings, and they have a taped report on the boat. Now I don't imagine if there was anything seriously wrong, they'd have that on the tape ... But can you imagine all of us calling every time there is a storm? (Wives Interviews)

Community members, wives, and families know that the North Atlantic fishery is a dangerous industry and that the men take the risks in order to provide for their families. Davis (1985) argues that in the Newfoundland context, "worrying" provides evidence of being a "good wife." Thus wives take on the role of "worrier" within the culture, whereas a man who worries is said to be an "old wife."

Fishers distinguish between "worrying" and "being in danger." When we asked fishers if their families worried about them during their absence at sea, all crew members and two-thirds of the officers including captains and mates said yes. Yet when we asked if their families thought their work dangerous, only 75 percent of the crew, 68 percent of the officers, and 63 percent of the captains and mates answered yes. A number of those men went on to explain how silly they felt their families were to hold those beliefs. Conversely, a number of officers explained that their families had once thought fishing dangerous, but that they had persuaded their families that it was not. An alternative hypothesis might be that once a crew member moves off the deck and into the wheel-house his wife feels more secure about his safety than when he worked on deck, but she continues to worry. Two wives we interviewed spoke of their relief when their husbands moved into the wheel-house. They recognized the difference in the level of danger and acted appropriately. This change in attitude by the family may be perceived by the fisher as a sign of acceptance of his work, rather than a sign of relief.

The family itself may also play a part in desensitizing individual members to the risks at sea. Poggie (1980, 125) found that the more relatives a fisher had in fishing the more satisfied he was with safety at sea. Thus fishing families may adapt to the psychological stress of fishing through family exposure to the risks and through positive role models. By the time children of fishers consider a career in fishing their father has become firmly established in the industry – after many years unsuccessful workers will have left the fishery and only the "winners" will remain. Since most learning at the initial stages occurs on an individual and informal basis, the majority of fishers' sons entering the industry will be initiated by their fathers, who have already made the psychological adjustments necessary to cope in this environment. But this socialization makes it more difficult to change attitudes to safety.

MAKING A DIFFERENCE

Workers' characterization of a situation determines the way they define risk. Unions, worker groups, companies, government agencies, and to some extent families who want to increase fishers' consciousness of the importance of safety must work not only to improve safety conditions on vessels but also to redefine fishers' attitudes and broaden their understanding of what factors influence safety, especially those factors outside the immediate workplace (e.g., government regulations and companies' management policies). Moreover, they must use an approach that presents information on how to reduce risks without undermining the perception of being in control, an essential component of many fishers' psychological well being. For example extraordinary situations, such as fires at sea, must be made ordinary. Monthly fire drill training in the actual conditions on board will allow fishers to feel in control of such a situation should it arise.

Companies, unions, and worker groups need to develop a positive approach to safety issues and to work together to enhance safety awareness. Demonstrations of safety equipment, fire and abandon-ship drills, and safety inspections need to be held frequently. Compulsory training programs must begin early in fishers' careers and those courses need to be supported by companies. Workers entering the fishery must be made aware of safety procedures from the beginning. A basic training course with safety awareness and first aid components needs to be developed. This class should be taken by greenhorns prior to or in their first three months of employment, as a condition of employment. Companies should recognize the safety advantages of a consistent crew membership and endeavour to keep crews together. Because fishing families play a crucial role in supporting current fishers and in socializing the next generation of fishers, attempts should be made – e.g., through public demonstrations of safety equipment – to draw them into the process.

CHAPTER TEN

Toward a Safer Fishery

The deep sea fishery works far out of the sight of land. Each large fishing vessel must be a self-contained and self-sustaining harvester of fish. Over the last decade technological improvements such as fish finders, de-icers, and boxing of fish have enhanced the ability of fishers to find, catch, and store fish. These technological advances mean better-quality fish, better dockside prices, and shorter trips. But well-designed and technologically advanced equipment should take safety and ergonomic factors, as well as economic ones, into account. A more efficient and safer workplace should in turn minimize accidents and decrease physical and economic risks. Vessels designed to work in bad weather and to fish in winter have equipment such as ice davits, de-icers, and reinforced hulls to cope with the icy waters, and FAX machines, and other electronic gear to warn of bad weather. Enhanced safety equipment such as inflatable life rafts, survival suits, EPIRBS (emergency position-indicating radio beacons) and other locating devices, on-board fire-fighting equipment, and government rescue services gives a greater range of help in emergencies. Thus recent technological improvements have both enhanced safety and working conditions and increased profits.

Yet this improved technology has a downside. First, the new equipment requires additional training of fishers and each new technology comes with specific safety concerns. For example, when boxing replaced the pen system, accident patterns associated with storing fish changed dramatically. The new risks must be identified and means of reducing them must be developed.

Second, younger crew members and officers, especially those who have spent a minimum of time on the deck, rely almost exclusively on current technology. If equipment breaks down, then traditional methods must be used; however, this new generation of fishers has not been taught those old skills. For example, few captains or mates can navigate by the stars or use a sexton if the lorans breaks. Thus the technology itself increases the crew's level of risk.

Third, some fishers misuse safety equipment. They take greater and greater risks by relying more and more on safety equipment to protect them from dangers. For example, they push their vessel and equipment to the limit in bad weather conditions rather than returning to shore. This cavalier attitude in dangerous situations may result in accidents and possibly death.

But sometimes a trade-off has to be made among general health concerns, improved technology, and safety equipment. For example, general health (the reduction of the risk of developing a chronic problem) might be improved at the cost of day-to-day safety in the working environment. To illustrate, earplugs protect workers on the deck or in the fishrooms from hearing impairment, but few fishers wear them. Those cramped workplaces full of machinery have restricted visibility and, because they are on the water, are unstable. Things happen quickly. Workers rely on each other for information on the stage of the work process, for instructions, and for warnings. Fishers wearing earplugs cannot hear their fellow workers call out instructions or warnings. Thus increased hearing over a lifetime comes at the price of reduced day-to-day safety.

MOVING TOWARD A WIDER PERSPECTIVE ON SAFETY

Studies on occupational health and safety in the fishery have usually focused on the immediate workplace. The suggested remedies have been improved technology with more safety or ergonomic features built in, improved safety equipment, such as better fire-fighting equipment, life-rafts, and survival suits, and the retraining of workers and reorganization of work in order to use those new technologies properly. This assumes that a controlled environment represents a safe environment, therefore more control equals improved safety. However, the work environment on a fishing vessel remains inherently uncontrollable. Conditions at sea can be predicted through weather forecasts, etc., but can never be controlled. Moreover, fishing remains essentially hunting a mobile resource at sea. No one knows when they will find fish, or what kind of fish, or how many, although this can be predicted to some extent through fish-finders, transducers, or other electronic equipment. Once found, the fish must be harvested immediately. Thus the harvesting sector will always be plagued with some uncertainty.

There must be an acceptance of this uncertainty, and planning must take into account this essential characteristic of the work. Many accidents happen at sea in bad weather. During a major storm, rescue services can make little difference. Helicopter or vessel transfer at sea cannot occur in gales. Safety programs should recognize this and put resources into developing such aids as medical emergency radio links between fishers on vessels and doctors and other medical personnel. Deep sea fishing boats should have better medical facilities and computer programs to consult when radio contact breaks down. Crew members need enhanced training in medical procedures; perhaps the cook should be trained as a paramedic or at least have a higher level of first-aid training.

But focusing on the immediate environment is not enough. I am not saying that such improvements and programs should be ignored: on the contrary, they remain essential. However, most of these remedies react to and correct problems with the technology in place. Programs must be proactive as well as reactive. As the industry adapts to external pressures and revises its ways of doing business it must assess the impact of the resulting technological changes on the workplace. As technology changes the physical environment of the workplace changes; the organization of work and the nonmonetary or extrinsic benefits of work change. Those changes affect safety, not only directly but indirectly. Workers weigh the benefits of such changes. What may appear to the outside observer an obvious benefit can be perceived by the worker as a nonbenefit.

Stricter safety regulations adapted from labour conditions on shore have decreased danger at sea. But the implementation of land-based regulations does not appear to be the answer either. For example, the right to refuse unsafe work has been essential to eliminating unsafe working conditions on land. It could be used to stop a number of unsafe work practices at sea, such as chaining off the warp or shooting and hauling back in extreme weather conditions. But does this make sense in all situations at sea? In some circumstances the refusal to work in unsafe conditions remains impossible. In those working conditions, the crew is literally in the same boat. When unsafe conditions arise they cannot escape, and must work cooperatively in order to survive. If the foredeck ices up, the ice must be removed or the vessel will sink. To go out and knock off the ice entails dangerous work in a dangerous environment. Yet someone must take the risk in order for the vessel and the rest of the crew to survive. What would happen if everyone refused to work in those conditions? The solution for the offshore fishery is not to import land-based answers but to tailor their own.

AFTER THE MORATORIUM

Over the last three years government biologists, fishermen, and company officials have come to realize that the North Atlantic fish stocks, in particular the northern cod, have been severely depleted. Many reasons have been given for this decline: overfishing by domestic and foreign fleets, mismanagement of the stocks by government officials, counter-productive harvesting practices such as high grading (the practice of discarding low-quality fish), increased predators (particularly gray and harp seals), increased morbidity of cod due to the increase of cod worms, and the deterioration of ocean conditions (particularly the lowered sea-water temperature). In February 1992 the federal government halted deep sea harvesting of northern cod off the northern shores of Newfoundland. In June of the same year the government expanded its moratorium to include inshore fishers. It subsequently issued a groundfish management plan, limiting all fishing of northern cod off Newfoundland as well as in specific areas of the Gulf of St. Lawrence, including Sydney Bight, and in waters off Nova Scotia north of Halifax. On 31 August 1993 the government closed five more areas, virtually halting all cod fishing off the Atlantic Canada shore north of Halifax except for fisheries off the Labrador coast. Groundfish quotas in the areas remaining open were severely cut.

According to Ken Kelly of the *National Fisherman* (1993, 14–15), 40,000 fishery jobs have been lost since the first northern cod closures. Although the federal government has promised to compensate fishermen for lost earnings and to offer them retraining programs, as well as early retirement packages for plant workers, the future for those men, their families, and their communities appears bleak. The compensation packages cannot go on forever, and even if the fishery rebounds it will not be able to support the large numbers of vessels and crews that previously exploited this resource. Moreover, in an area of high unemployment in the middle of a recession, many fishermen and plant workers are skeptical about what jobs they can retrain for and where those jobs will be located.

Many companies in Atlantic Canada have been forced to tie up their deep sea vessels during the summer months and to retire aging vessels. Many vessels now have two crews that rotate trips. For those men time on shore has increased from an average of two days to twelve or fourteen days and their wages have been cut in half. Some fish plants have been closed while others have lowered their capacity. Many seasonal plants have shortened their work time, and most year-round plants have extended their summer vacation period. These declines have inevitably had negative economic consequences for the local communities, and the government has not offered any relief to the local retailers or other people indirectly affected by the fishery closures.

However, in Nova Scotia decline, not moratorium, is the operative word. Particularly in the area south of Halifax and along the Fundy shore where the cod stocks remain less affected and the fishery is more diverse, the impact of the moratorium seems less severe than in other areas of Atlantic Canada. With the few fish available, the economic pressure in the deep sea fishery is even more intense. Questions of safety are just as pressing qualitatively, though fewer fishers risk their lives.

Severe storms and gales punctuated the cold and harsh winter of 1993. Yet the pressed deep sea companies continued to send vessels out to fish. In the early morning of 31 January 1993 the *Cape Aspey*, a scalloper, sank off Cape Sable Island while on her way to Georges Bank. The captain and four crew members drowned; the remaining eleven crew members survived. The interplay in this tragedy of depleted fish stocks, financial and market imperatives, bad weather, and attentiveness to safety has yet to be determined.

Notes

CHAPTER ONE

1 I use the term "fisher" to denote people who work in the fishery. There were no women working in the offshore fishery during my research period. I have retained the terms "fisherman" and "fishermen" in quotes and in the surveys.

2 Although women generally do not work in the offshore fishery, they do work in the inshore fishery and are starting to work in the midshore fishery. There is a group of women who work on midshore trawlers in Cape Breton. They have formed an association – Awareness of Women in the Fishery – to promote their interests.

3 The fieldworkers were coordinated by Jennifer Dingle, who worked as the first project manager. Much of the success of the survey is due to her administration.

4 Interviewers were usually able to contact either wives or mothers of fishermen to be interviewed while the men were out at sea, and find out when the men would be returning so that they could set up the interview. Single, separated, and divorced men not living in the parental home were more difficult to contact. This problem may have led to an underrepresentation of these men and to an additional bias to the sample.

5 All amounts are given in 1985 Canadian dollars, when one Canadian dollar was worth approximately eighty cents US, and represent before-tax incomes. The categories were (1) less than $10,000; (2) $10,000 to $24,999; (3) $25,000 to $49,999; (4) $50,000 to $74,000; and (5) $75,000 or more.

6 The best description of this technique comes from Babbie (1986, 263). He describes the technique as follows: "Snowball sampling is a method through which you develop an ever increasing set of sample observations.

You ask one participant in the event under study to recommend others for interviewing, and each of the subsequently interviewed participants is asked for further recommendations."

7 For a detailed discussion of the snowball technique and sampling procedures associated with achieving a probilistic sample see Biernacki and Waldorf (1981).

CHAPTER TWO

1 This discussion does not address the effects of those legislative changes. The legislation has apparently had little influence on the day-to-day workings of fishing vessels and their crews. The revisions of the Code shifted the responsibility for inspections of vessels and accident prevention from the provincial Workers' Compensation Boards to the Ship Safety Division of the Canadian Coast Guard. However, in the Atlantic region only one person, along with a half-time clerical worker for data collection, was assigned to this task. The forms used by fishers to register accidents and injuries at sea are being revised so that they more closely resemble the forms used at land-based work facilities, but at the time of writing the conversion has not yet been completed (Ship Safety Division personnel, personal communications, 1987–94).

2 "Atlantic Canada" or the Atlantic region includes the provinces of Nova Scotia, New Brunswick, Prince Edward Island, and Newfoundland. Thus companies based in Nova Scotia have fishing plants or facilities in Nova Scotia and may have one or more plants in the other provinces of the Atlantic region.

3 This plan set up Protection and Indemnity Associations ("P and I" Clubs) similar to ones in Britain and was associated with Lloyds of London. These insurance plans offered substantially less coverage than offered by the old workers' compensation plans. For example, hospital costs were not included. This meant that many workers had to carry additional insurance such as the Sick Mariners Fund. The Lunenburg Relief Society had been terminated once fishers were covered by the Workmen's Compensation Board and was not reintroduced when the board's coverage was suspended (Winsor 1987).

4 In 1972 Nova Scotia became the final province to include fishermen in its board's main program.

5 For an in-depth account of the UFAWU's attempt to organize the fishers and plant workers of Canso see Silver Donald Cameron (1977).

6 "Chaining off the warp" involves wrapping a chain around the warps (cables used for hauling in the net). This procedure keeps the warps in the midline of the ramp so that the warps lie in the vessel's ice-free wake.

CHAPTER THREE

1 The topics discussed in this chapter are also explored in Binkley (1989, 1990b).

2 For comparative material on other deep sea fishing fleets of the North Atlantic see Warner (1983).

3 The definitions of fishery type used here depend on statistical and licensing procedures of the Department of Fisheries and Oceans, and differ from those of other researchers. Some researchers maintain only the dichotomy of offshore and inshore while others such as Sinclair (1988, 3) maintain as many as four categories: inshore, nearshore, midshore, and offshore.

4 The share system gives a sense of egalitarianism to a group of highly independent men and thus allows for organization of the crew as a cooperative unit coming together to share the risks and profits of a joint venture.

5 For a general discussion of social and cultural characteristics of fishing people in marine environments see Pollnac (1988).

6 McGoodwin (1990, 8–12) makes a distinction between small- and large-scale fisheries based on level of capitalization. In his classification system both craft and industrial enterprises would be large-scale.

7 A few midshore vessels are family owned and operated. Fathers and sons or brothers may work side by side, but this is not the norm. The nature and physical demands of this type of work, coupled with the social costs of spending long periods at sea, puts too much strain on kinship ties (Thiessen and Davis, 1988, 12–13).

8 For a fuller treatment of recruitment practices in the community-based distant water fishery, see Thiessen and Davis (1988). For a discussion comparing kin-based recruitment with contract recruitment, see Stiles (1971). For an ethnographic account of a community-based fishery with both coastal and deep sea fisheries, see Davis (1985).

9 It is important to note that although the industrialized fishery is certainly becoming hierarchical it will never approach Weber's ideal type because of such features of the work organization as payment on a variable basis (share system or bonuses), need for skill, the dangers inherent in the occupation, and commitment to an occupation as opposed to an organization.

10 An important exception is the position of first mate, where masters (captains) may compete openly to hire a particular person.

11 The schedule of ten days on and forty-eight hours off typifies the experience of most trawler workers on vessels over one hundred feet. Those vessels fish all year round except for an annual refit, usually during the summer months. During the winter crews can be out for as long as

twelve to fifteen days and in the summer for as briefly as five to eight days. In smaller vessels of sixty-five to one hundred feet, the time on shore is usually slightly longer, but never longer than five days except for refit or for winter layover.

12 During the daily hail, in addition to instructions and transactions of company business, the company may relay personal information for crew members (such as a birth, illness, or death in a family).

13 In this specific case, rational management strategies and the drive for improved quality are integral. We have treated them as a single phenomenon.

14 Industrial fishing enterprises are characterized by rational management, a type of management that implies maximum control and large corporate industrial organization. That does not simply mean that there are managers; this system of management allows managers who are not present on the vessels to control and manage the company. Differences in management practices are reflected in the organization of the work. This includes recruitment practices, character and pace of the work, sense of personal control, method of payment, and the integration of workers into their home community. For more discussion on the characteristics of rational management see Cassin (1990) and Chandler (1977).

15 In this case the change from quantity to quality was rational. In other circumstances it could have been rational to opt for quantity over quality of fish caught, as it was before the enterprise allocation came into effect.

16 Under the *British North America Act*, 1867 (now referred to as the *Constitution Act*, 1867) Canada's federal government was granted jurisdiction over the fisheries. The Department of Fisheries and Oceans is the federal department that oversees and enforces fishery legislation, including the enterprise allocation.

17 Ironically, in 1992 this fishery was closed for two years because of the poor stocks available at that time.

18 The northern cod stocks were closed in Feburary 1992 for the deep sea fishery, but only one small company had any substantial quota left. It was saving its allocation for the better-quality fall cod. In July, seven months into a three-year management plan, the federal government modified the haddock/cod quota for the eastern stocks. Most deep sea companies had already used most of their quota. The government had to back down on fully implementing the quota reduction because of pressure from the affected companies, which argued that the stock was theirs.

19 The following description is a composite based on several trips that I took on deep sea vessels during the study period. Although this is not an actual trip, trips similar to this one often take place. This composite has been produced in order to maintain confidentiality and anonymity

of individuals I have sailed with. I have tried to preserve the characteristics of working conditions as faithfully as possible.

CHAPTER FOUR

1 The topics discussed in this chapter were also explored in Binkley and Thiessen (1988).

2 The controversy over having women aboard factory freezer trawlers e.g., on the *Cape North* to process the fish, is the exception that proves the rule. As the report on the socio-economic impact of the *Cape North* (Gardner Pinfold Consulting Economists 1987) indicates, it has taken unusual efforts by the company involved to break a community taboo of not having women at sea. High crew turnover, low profits, and community opposition from both sexes show the strength of the tradition. The taboo is based on fears for sexual and family stability.

3 Although fishing ventures are considered dangerous, and indeed are dangerous in Nova Scotia's deep sea fisheries, objective danger is not a necessary condition for worrying. For example, of a village in Peru, Degrys (1974, 40) notes that "according to the historical record, it appears that there has been only one death and no serious injuries at sea in the last seventy or eighty years." Yet the villagers "declare the extreme dangers of fishing as an occupation."

4 It should be noted that bingo and lotteries are seen as magical solutions that will not only end the financial problems but also relieve the boredom of everyday life, in a socially acceptable way.

5 In a Peruvian fishing community where visual sighting of the vessel is the only warning, the women are expected to be at the beach when the boat lands (Degrys 1974). There is a lapse of only about five minutes between sighting and landing. She remarks that "If a woman does not meet her husband on time, serious domestic trouble follows" (ibid., 41).

6 A disproportionate number of migrant fishers' wives are also clients of these centres. Migrant women have no family support group to turn to for help, nor is there any family group encouraging the couple to remain together. Thus the migrant wives have more "freedom of action" than local women, but they also have less protection.

7 I would like to thank Victor Thiessen for the development of this argument.

CHAPTER FIVE

1 The topics discussed this chapter were also explored in Binkley (1990b).

2 In this chapter, I chose to use descriptive categories such as "monetary" and "nonmonetary" because those terms arose during fieldwork. The expression of workers – the way they describe and understand phenomena,

and their relation to it – is more important than the characterization of that expression. Of course those phenomena could be characterized in other ways, for example, as intrinsic and extrinsic benefits (cf. Burrell and Morgan 1979).

3 Since job satisfaction is a multifaceted phenomenon, Poggie and Pollnac used a number of measures. These included two general measures of job satisfaction and a battery of twenty-six attitudinal questions using a scale from 1 to 5 to measure variation in satisfaction. Through the use of factor analysis, they identified underlying factors affecting levels of job satisfaction.

4 Gatewood and McCay used four job designations – captain, first mate, crew, and one-man – rather than just "captain" versus "crew". The "one-man" designation was used because bay-fishing is a one-man operation. Most of their analyses focused on status, contrasting captain, mate, and crew from the five fisheries where such distinctions are relevant, and they left out baymen (McCay, personal communication, March 1987).

5 I did not ask, "If you had a daughter would you want her to go fishing?" As mentioned earlier, Nova Scotian women do not go to sea on offshore boats as fishers, although they do "help out" on inshore vessels or occasionally work on the fish processing line on the factory freezer trawler. If I had asked the men I was interviewing this question my credibility as a sensible person would have been in doubt. In the unstructured interviews I did ask fishers who had daughters if they would like their daughter to marry a fisherman. Invariably they said it was up to her, but it was a hard life.

6 Apostle, Kasdan, and Hanson (1985) and Binkley (1990b) used factor analysis on these job satisfaction items. There is considerable similarity between their dimensions and the indices that I will develop. The main difference is that safety/health considerations did not emerge as a factor in their studies; it is clearly an important domain for us. Also, in the context of this study, the factor Work Quality is better renamed Stress.

7 When using Cronbach's alpha to test for internal consistency and reliability, a value of .60 is acceptable and a value of .70 is good (Marsh 1977, 254–5; cf. Apostle, Kasdan, and Hanson 1985, 284).

8 This is a general problem in retrospective interviews. Past methodological studies have repeatedly indicated that for most topics recollections are considerably biased. Respondents may be willing to tell a researcher what they thought about something that happened several years ago, but the responses are usually poor reflections of what they actually thought at the time about the topic.

9 A principal component analysis yielded results similar to the factor analysis for the entire structure.

10 "Sometimes these values are regarded as an estimate of the relative importance of factors, but one must be cautious about such a simple view, as variations in eigenvalues may merely reflect a difference in the number of potential questions which measure a particular factor. Such an imbalance in the number of questions may well explain why the first factor has appeared first, since there could be more questions loading on it at the 0.4 level" (Apostle, Kasdan, and Hanson 1985, 285).

11 This factor was identified by Apostle, Kasdan, and Hanson (1985) as Crowding, but this is a somewhat ambiguous label. It would be more appropriately called Acceptable Level of Crowding.

12 For a more detailed discussion of the effects of job satisfaction on recruitment and attrition of workers in the deep sea fishery, see Binkley (1989), and Binkley and Thiessen (1990).

13 The most notable of those earlier studies are Aubert and Arner (1958); Horbulewicz (1972); Nolan (1973); Tunstall (1962).

14 Similar findings in relation to size of enterprise have been reported in other industries that are going through the process of industrialization.

15 The only exception was Gatewood and McCay's (1988) finding that satisfaction with government officials was significantly higher among oystermen. This fishery is not represented in our study.

16 "Turnover status" refers to whether a respondent is a former or a current deep sea fisher. In other words, this variable simply indicates the sample – stayers or leavers – to which the respondent belongs.

17 It is of course possible to compute these similarity profiles using other measures of association such as Spearman's rank order coefficients. As its name implies, this procedure converts the means into ranks from 1 to 26 for each of the six groups and then correlates the ranks. We, like Bohrnstedt and Borgatta (1985), consider the assumptions made in this "non-parametric" technique to be at least as problematic as the assumptions of normality and interval measures made when computing Pearson's correlation coefficients. In particular, the Spearman's procedure forces a minute difference in two adjacent means to be given the same weight as a large difference between another adjacent pair of means. Since our scores are means, it is reasonable to give more weight to large differences than to small ones. Spearman's correlation coefficients were nevertheless computed with mixed results. That is, the patterns reported in table 5.11 for the Pearson's correlations are also manifest when Spearman's r's are computed, but less clearly.

CHAPTER SIX

1 "Highliner" is a term applied to a captain, crew, or vessel that is very successful at catching fish.

2 For a more thorough discussion of the type of competition that takes place between skippers, see Tunstall (1962); Andersen (1979, 1972); Andersen and Stiles (1973).

3 For further discussion on these issues, see Davis and Thiessen (1986) and Binkley (1990b).

4 John Herbin, Nova Scotia Department of Labour, (personal communications, 1986–87). Also, see Neis (1986, Appendices A and B) for a general discussion on health and safety concerns related to technological change in the Canadian east coast fishery.

5 See Binkley and Thiessen (1990) for an earlier discussion of this topic.

6 For a fuller discussion of the implications of the enterprise allocation and its concomitant management practices on working conditions in the deep sea fishery see Binkley (1989).

7 For a detailed discussion of how this trade-off is made by a group of New England fishers see Poggie (1980, 127).

CHAPTER SEVEN

1 Labour Canada has not yet generated its own statistics concerning marine workers.

2 B. Thorne (personal communication, 1986). An alternative way to analyse the mortality of active fishers has been demonstrated by Dr. I. Neutal (personal communications, 1987, 1988). Using demographic characteristics, Dr. Neutal linked Department of Fisheries and Oceans (DFO) licensing data (1975, 1977–83) to Statistics Canada mortality data (1975–83). This method has a number of shortcomings, however. Most notably, death certificates are issued only when a body has been examined. This means that a death at sea where the body is not recovered is not recorded on a death certificate and is not included in the Statistics Canada data, which are based on death certificates. This has led to an underreporting of deaths due to drownings, suicides, and other accidents at sea where the body is not recovered. The DFO material also has deficiencies because the definition of "licensed fisher" has changed over the study period, and the data for 1976 were excluded because DFO did not enter the licensing data for that year in computer-readable form. Relative to using published government statistics, this method is time-consuming, expensive, and yields little information beyond the direct cause of death and the demographic characteristics of the individual.

3 The categories presented on the precoded form are the same categories used in the tables generated from this information. There is little opportunity to deviate from these official categories or to develop new

categories because of the structured nature of the form. However, people filling out the form may interpret the categories differently, and some variables may be introduced into the information.

4 This category reflects a historic bias; it goes back to when malnutrition was common among sailors and other seagoing workers. Malnutrition is still an important problem for some foreign fleets.

5 The agency recommends modification of safety legislation in order to eliminate or reduce those extraordinary incidents. One of the ironies of the situation is that although the legislation may lead to safer vessels with better life-saving equipment it may also lead to owner/employers pushing their vessels beyond safe limits by fishing in more extreme conditions in order to recoup the costs of making their vessels safer.

6 Fifty-six percent of the sample sought medical aid for acute work-related conditions – respiratory and gastric complaints, infections, cuts, fractures, and muscle sprains and strains. About 21 percent went for treatment of either chronic ailments associated with past injuries or arthritis. Both groups are included in the number of those who sought medical aid.

CHAPTER EIGHT

1 These figures were statistically significant, with a chi-square value of 12.36 with 2 degrees of freedom and a probability level of .002.

2 A finding that was statistically significant at ($p = .006$) using a chi-square test with Pearson value of 10.111 with two degrees of freedom.

3 Neutal (1989) reports that fishers have a lower rate of smoking-related diseases than other workers. She argues that this may be due to healthy worker effect. However, it might also be due to systematic exclusion of fishers with those problems from her sampling, thus rendering the finding an artefact.

4 Cigarette packs come in three sizes: large, with twenty-five cigarettes: regular, with twenty cigarettes; and economy (nicknamed the "poverty pack"), with fifteen. I chose the twenty-five cigarette pack because it is the most common size and is commonly used in comparative studies.

5 Rix, Hunter, and Colley (1982) argue that drinking before a trip minimizes fishers' anxiety about leaving their homes and families.

6 The work schedule of fishers does not allow fishers' drinking patterns to be compared with those of other blue-collar workers. This is one of the reasons why fishers are not included in statistical compilations of high-risk drinkers.

7 The chi-square value (20.79) is statistically significant at the probability level of greater than .001 with four degrees of freedom.

8 Using *t*-test the comparison of captains/mates vs other officers was statistically significant at the $p = .04$ level (2–tailed) for *f*-test value of 2.06 with 162 degrees of freedom.

9 The term was coined by Kline, Robbins and Thomas in their 1989 article.

10 Riordan, Johnson, and Thomas (1991, 391) argued that for fishers "a greater sense of mastery was associated with reduced stress."

11 Schilling (1966) saw these traits as stoic acceptance of extremely dangerous conditions.

12 There is a belief that many men who drown at sea are committing suicide.

CHAPTER NINE

1 Poggie (1980, 125) found that "the more kinsmen a person had who are fishermen the more satisfied the individual was with the safety of his work." Thus I would expect this variable to have a positive effect on Nova Scotia fishers' satisfaction with safety. I did not ask fishers how many kinsmen they had who were fishers, so the hypothesis cannot be tested.

2 In their earlier paper, Pollnac and Poggie identified the possibility of interplay between "age" and "years of experience." Following their example, "age" and "years of experience" were dichotomized at the sample means (39.2 years and 17 years respectively) and crosstabulated by the dichotomized version of the satisfaction scale. The findings for the two variables are virtually the same. There is no statistically significant finding in this analysis; however, there is a tendency for younger and less experienced fishers to be less satisfied than the older and more experienced workers with safety. Pollnac and Poggie (1989, 4) identified a similar trend for the "years of experience" variable.

3 These three additional factors are specific to the conditions of the Nova Scotia deep sea fishery. They do not apply to the New England fishery, where safety courses and corporate interest in safety are not common.

4 Analysis of the relationship of level of satisfaction with safety to the variables "job status", "enterprise type," and "sector of the fishery" did not yield statistically significant results, but two related trends should be noted. First, crew members were less satisfied with safety than officers and captains were. Second, fishers working in craft enterprises (the midshore fishery) were more satisfied with safety than workers in industrial enterprises (the deep sea sector) were. When the relationship between the general job satisfaction variables – "Would you go fishing again?" and "Do you want your son to go fishing?" – and the level of satisfaction with safety were analysed, the results were not statistically significant.

Note, however, that those workers who stated that they would not advise a son to go fishing or would leave it up to him were proportionately more dissatisfied with the present levels of safety than were those who would advise their sons to go fishing.

5 This analysis was significant with a F ratio of 10.2 at the p = .00 level with two degrees of freedom.

6 There may be some interplay between "company attitude" and "vessel type". National Sea is the most progressive company and the majority of its vessels are trawlers.

7 The situation is the reverse in the Newfoundland deep sea fishery. There the driving forces for change are the unions – UFCW and NFAW – not the companies. For more information about the Newfoundland situation see Barbara Neis (1986).

8 See chapter 8 regarding the responses to the question, "Have you ever had a serious accident?"

Glossary

Bank: An area of the sea bottom that rises significantly above the surrounding ground.

Bayman: An inshore fisherman who traditionally fished in the local bay.

Block: A pulley used to alter the direction of a rope or chain. Two or more blocks constitute a tackle.

Bottom trawling: Towing nets over the sea bed on rollers, as in cod fishing.

Break deck: The midship end of the raised deck.

Bridle: Two or more parts of a rope or chain rigged to distribute stress on another rope or chain to which it is connected.

Broker: A fishing trip on which the fishers break even or lose money. (Also a fish broker, an intermediary who buys the fish from the company or the fishermen.)

Chance: Giving a man a chance means giving him a job fishing.

Choke: To foul a rope in a block. To choke the luff means to jam the hauling part of the tackle to prevent it from stacking up.

Clubstick: The metal bar at the lower end of the scallop rack net where the fish collect, equivalent to the *cod end* on a trawl.

Cod end: The lower end of the fishing net where the fish collect (see figure 3.4).

Davit: A crane on board a vessel.

Doors: See *otter trawl.*

Dragger: Any fishing vessel, such as a trawler or scalloper, that drags fishing gear through the water.

Dump chains: Chains at either end of the dump table that lift that portion of the deck to dump debris overboard.

Dump table: The part of the deck on a scalloper where the rake drops its contents, which then lifts to dump debris (rocks and mud) overboard after the scallops have been collected.

EA (enterprise allocation): An individual quota for a specific species assigned by the federal government to an offshore, vertically integrated fishing enterprise. EAS are not transferable.

EPIRB (emergency position-indicating radio beacon): A locating device on a vessel that is activated when the vessel is in distress.

Gallows: A horseshoe-shaped girder hinged on the deck of a trawler. It carries the *block* through which the *trawl warp* is led.

Gear: Tackle, ropes, *blocks*. Any working equipment for shipboard operation.

Groundfish: Any fish living at or near the sea bottom, such as halibut, cod, or flounder.

Hauling back: Pulling in the *cod end*, belly, and wings of the net, which are being dragged behind the vessel, over the stern while the vessel moves forward and hauls in the *trawl warps* through the winch.

Hawser: A flexible steel wire or fibre rope, used for hauling, warping, or mooring.

Highliner: A captain, crew, or vessel that is very successful at catching fish.

ITQ (individual transferable quota): An individual quota for a specific species assigned by the federal government to an inshore or midshore fishing vessel. These are transferable.

Knock-out block: The block that keeps a *scallop rake* suspended. When it is removed the rake falls into the water.

Longliner: An inshore or midshore vessel that fishes using longlines – long lines with hooks at intervals. Longlines either float or are sunk and anchored on the bottom. They can be used in any kind of weather and where fish are too scattered for effective trawling.

Lorans (long range navigational system): A system of navigation using signal pulses from radio transmitters.

Midwater trawl: A towed net that can be set at different depths. It is used for redfish, herring, cod, capelin, dogfish, squid, cusk, silver hake, argentine, and shrimp.

Otter trawl: Any towed net that has a pair of otter boards. An otter board is a board rigged with a line and *bridle* which, when lowered underwater, stretches the line by reason of the outward angle at which the bridle holds it. Used to spread out the net or to keep open the mouth of the trawl. Otter boards are also called *doors* on trawl nets.

Scallop rake: A large rake towed over the seabed. See figures 3.7 and 3.8.

Seining, or purse seining: Fishing with seine nets – nets that encircle schools of fish (tuna, herring, mackerel). Purse seining cannot be done in as rough weather as other types of fishing.

Shooting away, or shooting the gear: Throwing the cod end, belly, and wings of the net, which have been stowed along the rail, out over the stern while the vessel moves forward and pays out ground cables from the winch.

Side trawler: A vessel from which nets are towed from over the sides of the boat.

Speed-up: Increasing the rate of work by lengthening normal shifts and shortening rest periods to take advantage of good fishing.

Stern trawler: The most efficient deep sea fish catcher, with a crew of approximately fourteen, making trips of twelve to fourteen days, capable of carrying up to 400,000 lbs of iced groundfish.

Surfclam: A species of clams found on the near shore.

Ticket: A Certificate of Competency issued by the federal Department of Transport to a seaman who has passed an exam in a specified grade and has been found fit to perform the duties of the grade.

Transducer: A depth finder sewn into the top of a fishing net. It is used in midwater trawling to control the depth of the net.

Trawl balls: Steel balls attached to the bottom of the trawl net that allow the net to roll along the bottom of the sea.

Trawl warp: The wire or rope leading through a block in the gallows by which a trawl net is towed.

Trawling: Towing large nets behind a boat.

Turnbuckle: A stretching screw designed to take up slack on metal wires.

Turnhook: A large metal hook, secured to the gallows by a wire strap, that is then hooked into the top of the scallop rake in order to hoist it up and dump the scallops on deck on the dump tables.

References

Andersen, Raoul. 1979. "Public and Private Access Management in Newfoundland Fishing." In *North Atlantic Maritime Cultures*, edited by Raoul Andersen, 299–336. The Hague: Mouton Press.

— 1972. "Hunt and Deceive: Information Management in Newfoundland Deep Sea Trawler Fishing." In *North Atlantic Fishermen*, edited by Raoul Andersen and Cato Wadel, 120–140. St. John's, Nfld.: Institute of Social and Economic Research.

Andersen, Raoul, and Geoffrey Stiles. 1973. "Resource Management and Spatial Competition in Newfoundland Fishing: An Exploratory Essay." In *Seafarer and Community*, edited by Peter Fricke, 44–66. London: Croom Helm.

Anderson, Lee G. 1980. "The Necessary Components of Economic Surplus in Fisheries Economics." *Canadian Journal of Fisheries and Aquatic Sciences* 37: 858–70.

Andro, M., Patrick Dorval, G. Le Bouar, Y. Le Roy, C. Roullot, and C. Le Pluart. 1984. *Les accidents du travail dans la peche maritime*. Partie III. Lorient, France: Institut universitaire de technologie de Lorient.

Anonymous. 1988. "Warning: How to Lose a Family." *National Fisherman*, September, 58.

Antler, Ellen Piles. 1981. "Fishermen, Fisherwomen, Rural Proletariate: Capitalist Commodity Production in the Newfoundland Fishery." Ph.D. diss., University of Connecticut.

Apostle, Richard, and Gene Barrett. 1992. *Emptying Their Nets: Small Capital and Rural Industrialization in the Nova Scotia Fishing Industry*. Toronto: University of Toronto Press.

Apostle, Richard, Leonard Kasdan, and Art Hanson. 1985. "Work Satisfaction and Community Attachment among Fishermen in Southwest Nova Scotia." *Canadian Journal of Fisheries and Aquatic Sciences* 42: 256–67.

Armstrong, Pat, and Hugh Armstrong. 1978. *The Double Ghetto; Canadian Women and Their Segregated Work*. Toronto: McClelland and Stewart.

Astrand, Irma, P. Fugell, C.G. Karlsson, Kare Rodahl, and Zdenek Vokac. 1973. "Energy output and work stress in coastal fishing." *Scandinavian Journal of Clinical Laboratory Investigations* 31: 105–13.

Aubert, Vilhelm, and Oddvar Arner. 1958. "On the Social Structure of the Ship." *Acta Sociologica* 3: 200–19.

Babbie, Earl R. 1986. *The Practice of Social Research.* 4th ed. Belmount, Calif.: Wadsworth Publishing Company.

Barney, W., and G. Carberry. 1987. *Offshore Scallop Fishing Experiment: St. Pierre Bank, 1985.* Canadian Technical Report of Fisheries and Aquatic Sciences 1548 St. John's, Nfld.: Department of Fisheries and Oceans, Fisheries Development Division.

Barrett, Michelle. 1980. *Women's Oppression Today: Problems in Marxist Feminist Analysis.* London: Verso Press.

Barrett, Michelle, and Mary MacIntosh. 1983. *The Anti-Social Family.* London: Verso Press.

Beechey, Veronica. 1987. *Unequal Work.* London: Verso Press.

Bem, Daryl J. 1972. "Self-Perception Theory." *Advances in Experimental Social Psychology* 6: 1–62.

Bergman, David. 1978. *Death on the Job: Occupational Health and Safety Struggles in the U.S.* New York: Monthly Review Press.

Biernacki, P., and D. Waldorf. 1981. "Snowball Sampling: Problems and Techniques of Chain Referral Sampling." *Sociological Methods and Research* 10: 141–63.

Binkley, Marian. 1994. *Voices from Off Shore.* St. John's, Nfld.: ISER Books.

– 1991. "Nova Scotian Offshore Fishermens' Awareness of Safety." *Marine Policy* 15(3): 170–182.

– 1990a. "The Concept of Risk in the Modern Fisheries – Physical, Economic and Social Considerations." In *The Proceedings of the First International Symposium on Safety and Working Conditions aboard Fishing Vessels,* edited by J.P. Roger, 119–22. Rimouski, Quebec: University of Quebec at Rimouski.

– 1990b. "Work Organization among Nova Scotian Offshore Fishermen." *Human Organization* 49(4): 395–405.

– 1989. "Nova Scotian Offshore Groundfish Fishermen: Effects of the Enterprise Allocation and the Drive for Quality." *Marine Policy* 13(4): 274–84.

– 1986. "The Hidden Cost of the Offshore Fishery." Paper presented at a seminar on Research and Public Policy Formation in the Fisheries: Norwegian and Canadian Experiences. Tromsø, Norway: Institute of the Fisheries.

– 1985. "Impact of Changes in the Nova Scotia Fishery on Occupational Health and Safety of Trawlermen." Special Publication of *Canadian Journal of Fisheries and Aquatic Sciences* 72: 2–4.

Binkley, Marian, and Victor Thiessen. 1990. "Levels and Profiles of Job Satisfaction of Current and Former Offshore Fishers." *MAST* 3(1): 48–68.

- 1988. "Ten Days a 'Grass Widow' – Forty-eight Hours a Wife: Sexual Division of Labour in Trawlermen's Households." *Culture* 8(2): 39–50.

Bohrnstedt, George W., and Edgar F. Borgatta, eds. 1981. *Social Measurement: Current Issues.* Beverly Hills, Calif.: Sage.

Bourne, N. 1964. *Scallops and the Offshore Fishery of the Maritimes.* Bulletin 145. Ottawa: Fisheries Research Board of Canada.

Braverman, Harry. 1975. *Labor and Monopoly Capital: The Degradation of Work in the Twentieth Century.* New York: Monthly Review Press.

Bray, G.A. 1982. *Obesity and Overweight Study Guide.* Washington: American Medical Association.

Burawoy, Michael. 1985. *The Politics of Production: Factory Regimes under Capitalism and Socialism.* London: Verso Press.

- 1972. *The Colour of Class on Copper Mines, from African Advancement to Zambianization.* Manchester: Manchester Press for the Institute for African Studies, University of Zambia.

Burns, Tom, ed. 1969. *Industrial Man.* London: Penguin Books.

Burrell, Gibson, and Gareth Morgan. 1979. *Social Paradigms and Organisational Analysis.* London: Heinemann.

Cameron, Silver Donald. 1977. *The Education of Everett Richardson.* Toronto: McClelland and Stewart.

Canada. *Canada Labour Code. Marine Occupational Safety and Health Regulations,* SOR/87-183, am. SOR/88-198.

- *Canada Shipping Act. Shipping Casualties Reporting Regulations,* C.R.C. 1978, c. 1478, rep. & sub. SOR/85-514; am. SOR/85-1088.

Canada. Committee on Occupational Safety and Health in the Fishing Industry. 1988. *Report.* Ottawa: Labour Canada.

Canada. Department of Fisheries and Oceans. 1987. *Social and Economic Impact of the Factory Freezer Trawler.* Ottawa: Department of Fisheries and Oceans.

- Atlantic Fisheries Service. 1984. *Licence Directory – 1983.* Ottawa: Department of Fisheries and Oceans.

- Economic Analysis and Statistics Division. Economic and Commercial Analysis Directorate. 1991. *Canadian Fisheries Annual Statistical Review 1986.* Canadian Fisheries Annual Statistical Review, vol. 19. Ottawa: Department of Fisheries and Oceans.

Canada. Health and Welfare Canada. 1986. *Smoking Behaviour of Canadians, 1986.* Ottawa: Health and Welfare Canada.

Canada. Labour Canada. 1986. *Employment Injuries and Occupational Illnesses, 1981–1984.* Ottawa: Labour Canada.

- 1983. *Employment Injuries and Occupational Illnesses, 1972–1981.* Ottawa: Labour Canada.

Canada. Royal Commission to Investigate the Fisheries of the Maritimes and the Magdalen Islands. 1928. *Report.* Ottawa: F.A. Acland, Printer.

Canada. Task Force on Atlantic Canada. 1983. *Navigating Troubled Waters: A New Policy for the Atlantic Fisheries.* (The Kirby Report.) Ottawa: Supply and Services Canada.

Canada. Transport Canada. Canadian Coast Guard. Coast Guard Working Group on Fishing Vessel Safety. 1987. *A Coast Guard Study into Fishing Vessel Safety.* Ottawa: Transport Canada.

– Marine Casualty Investigations. 1976–82. *Marine Casualties System, Annual Public List.* Ottawa: Transport Canada.

– Marine Casualty Investigations. 1982–91. *Statistical Summary of Marine Accidents, 1981–90.* Ottawa: Transport Canada.

Cassin, A. Marguerite. 1990. "The Routine Production of Inequality: A Study in the Social Organization of Knowledge." Ph.D. diss., University of Toronto.

Chambres, Larry W., and Walter O. Spitzer. 1977. "A Method of Estimating Risk for Occupational Factors Using Multiple Data Sources: The Newfoundland Lip Cancer Study." *American Journal of Public Health* 67: 176–9.

Chandler, Alfred D. Jr. 1977. *The Visible Hand: The Managerial Revolution in American Business.* Cambridge, Mass.: Harvard University Press.

Child, John. 1984. *Organizations: A Guide to Problems and Practices.* London: Harper and Row.

– 1975. "Managerial and Organizational Factors Associated with Company Performance – Part II: A Contingency Analysis." *Journal of Management Studies* 12(1): 12–27.

Child, John, and Bruce Partridge. 1982. *Lost Managers: Supervisors in Industry and Society.* Cambridge: Cambridge University Press.

Clegg, Stewart, and David Dunkley. 1980. *Organization, Class and Control.* London: Routledge and Kegan Paul.

Clement, Wallace. 1981. *Hardrock Mining: Industrial Relations and Technological Changes at Inco.* Toronto: McClelland and Stewart.

Collacott, R.A. 1977. "Risks to Trawler Fishermen in Orkney Waters." *Journal of the Royal College of General Practitioners* 27: 482–5.

Connolly, Patricia M. 1978. *Last Hired, First Fired: Women and the Canadian Workforce.* Toronto: Women's Press.

Connolly, Patricia M., and Martha MacDonald. 1985. "Women and Development: The More Things Change, the More They Stay the Same." In *Women and Offshore Oil,* 392–428. Conference Papers no. 2. St. John's, Nfld.: Institute of Social and Economic Research.

Conway, Terry L., Ross R. Vickers, Harold W. Ward, and Richard H. Rahe. 1981. "Occupational Stress and Variation in Cigarette, Coffee, and Alcohol Consumption." *Journal of Health and Social Behavior* 22(2): 155–65.

Cooper, M. Lynne, Marcia Russell, and Michael R. Frone. 1990. "Work Stress and Alcohol Effects: A Test of Stress-Induced Drinking." *Journal of Health and Social Behavior* 31(3): 260–76.

Crawford, Robert. 1977. "You Are Dangerous to Your Health: The Ideology and Politics of Victim Blaming." *International Journal of Health Services* 7(4): 663–80.

Danowski, Fran. 1980. *Fishermen's Wives: Coping with an Extraordinary Occupation.* Marine Bulletin no. 37. Kingston, R.I.: International Center for Marine Resource Development.

Davis, Anthony. 1985. " 'You're Your Own Boss': An Economic Anthropology of Small Boat Fishing in Port Lameron Harbour, Southwestern Nova Scotia." Ph.D. diss., University of Toronto.

Davis, Anthony, and Victor Thiessen. 1986. "Making Sense of the Dollars: Income Distribution among Atlantic Canadian Fishermen and Public Policy." *Marine Policy* 10(3): 201–14.

Davis, Dona Lee. 1983. "Woman the Worrier: Confronting Feminist and Biomedical Archetypes of Stress." *Women's Studies* 10: 135–46.

Decker, Kathryn Brown. 1978. "Coping with Sea Duty: Problems Encountered and Resources Utilized during Periods of Family Separation." In *Military Families: Adaptation to Change,* edited by Edna J. Hunter and D. Stephen Nice, 113–29. New York: Praeger.

Degrys, Mary. 1974. "Women's Role in a North Coast Fishing Village in Peru: A Study of Male Dominance and Female Subordination." Ph.D. Diss., New School for Social Research.

Dixon, Richard D., Roger C. Lowery, James C. Sabella, and Marcus J. Hepburn. 1984. "Fishermen's Wives: A Case Study of a Middle Atlantic Coastal Fishing Community." *Sex Roles* 10(1/2): 33–52.

Douglas, Mary. 1986. *Risk Acceptability According to the Social Sciences.* London: Routledge and Kegan Paul.

Doyal, Leslie. 1979. *The Political Economy of Health.* London: Pluto Press.

Faris, James. 1966. *Cat Harbour: A Newfoundland Fishing Settlement.* St. John's, Newfoundland: ISER Books.

Festinger, Leon. 1957. *A Theory of Cognitive Dissonance.* Evanston, Ill.: Row, Peterson.

Fodor, J.G., C.J. Pfeiffer, and V.S. Papezik. 1973. "Relationship of Drinking Water Quality (Hardness–Softness) to Cardiovascular Mortality in Newfoundland." *Canadian Medical Association Journal* 108: 1369–73.

Gardner Pinfold Consulting Economists. 1987. *A Socio-Economic Impact Study of the Factory Freezer Trawler, "The Cape North."* Ottawa: Department of Fisheries and Oceans.

Gatewood, John B., and Bonnie McCay. 1990. "Comparison of Job Satisfaction in Six New Jersey Fisheries: Implications for Management." *Human Organization* 49(1): 14–25.

– 1988. "Job Satisfaction and the Culture of Fishing: A Comparison of Six New Jersey Fisheries." *MAST* 2: 103–28.

– 1986. "Comparison of Job Satisfaction in Six New Jersey Fisheries: Implications for Management." Paper presented at the annual meeting of the American Fisheries Society, Providence, R.I.

Gough, Ian. 1979. *The Political Economy of the Welfare State*. London: Macmillan.

Gray, R.J. 1987. *An Examination of the Occupational Safety and Health Situation in the Fishing Industry in B.C.* Ottawa: Labour Canada.

Hackman, J.R., J.L. Pearce, and Wolff, J.C. 1978. "Effects of Changes in Job Characteristics on Work Attitudes and Behaviour." *Organizational Behaviour and Human Performance* 21(3): 289–304.

Hingson, Ralph, Thomas Mangione, and Jane Barrett. 1981. "Job Characteristics and Drinking Practices in the Boston Metropolitan Area." *Journal of Studies on Alcohol* 42(9): 725–38.

Horbulewicz, Jan. 1972. "The Parameters of the Psychological Autonomy of Industrial Trawler Crews." In *Seafarer and Community*, edited by Peter Fricke, 67–84. London: Croom Helm.

Innis, Harold. 1954. *The Cod Fisheries: The History of an International Economy*. Toronto: University of Toronto Press.

Janes, Craig R., and Genevieve Ames. 1989. "Men, Blue Collar Work and Drinking: Alcohol Use in an Industrial Subculture." *Culture, Medicine and Psychiatry* 13(3): 245–74.

Jensen, L.B. 1980. *Fishermen of Nova Scotia*. Halifax: Petheric Press Ltd.

Kelley, Ken. 1993. "Atlantic Canada Reels under Fishing Closures." *National Fisherman*, December, 14–15.

Kelly, John E. 1982. *Scientific Management, Job Redesign, and Work Performance*. New York: Academic Press.

Kipling, Rudyard. 1982. *Captains Courageous*. 1897. Reprint, New York: Bantam Books.

Kline, Annette, Michael C. Robbins, and J. Stephen Thomas. 1989. "Smoking as an Occupational Adaptation among Shrimpfishermen." *Human Organization* 48(4): 351–4.

Knight, Rolf. 1971. *Work Camps and Single Enterprise Communities in Canada and the U.S.: A Working Bibliography*. Toronto: Scarborough College, University of Toronto.

Lacasse, François D. 1970. *Women at Home: The Cost to the Canadian Economy of the Withdrawal from the Labour Force of a Major Proportion of the Female Population*. Ottawa: Information Canada.

Leyton, Elliott. 1975. *Dying Hard: The Ravages of Industrial Carnage*. Toronto: McCelland and Stewart.

Lucas, Rex A. 1971. *Minetown, Milltown, Railtown: Life in Canadian Communities of Single Industry*. Toronto: University of Toronto Press.

Luxton, Meg. 1980. *More Than a Labour of Love*. Toronto: Women's Press.

Luxton, Meg, and Harriett Rosenberg. 1986. *Through the Kitchen Window: The Politics of Home and Family*. Toronto: Garamond Press.

MacDonald, J.M., and G.D. Powers. 1990. "United States Coast Guard's Recent Activities in Commercial Fishing Industry Vessel Safety." in *The Proceedings of the First International Symposium on Safety and Working Conditions aboard Fishing Vessels*, edited by J.P. Roger. Rimouski, Quebec: University of Quebec at Rimouski.

Marsh, Alan. 1977. *Protest and Political Consciousness*. Beverly Hills: Sage.

Maslow, Abraham H. 1954. *Motivation and Personality*. New York: Harper and Row.

McCubbin, Hamilton I., and Barbara B. Dahl. 1976. "Prolonged Family Separation in the Military: A Longitudinal Study." In *Families in the Military System*, edited by Hamilton McCubbin et al., 112–43. Beverly Hills: Sage.

McGoodwin, James R. 1990. *Crisis in the World's Fisheries: People, Problems and Policies*. Stanford: Stanford University Press.

Moore, Robert. 1985. "The Effects of Offshore Oil on Onshore Women." In *Women and Offshore Oil*, 392–428. Conference Papers no. 2. St. John's, Nfld.: Institute of Social and Economic Research.

Moore, S.R.W. 1969a. "The Mortality and Morbidity of Deep Sea Fishermen Sailing from Grimsby in One Year." *British Journal of Industrial Medicine* 26: 25–46.

– 1969b. "The Occupation of Trawl Fishing and the Medical Aid Available to the Grimsby Deep Sea Fishermen." *British Journal of Industrial Medicine* 26: 1–24.

Murray, Douglas L. 1982. "The Abolition of the Short Handled Hoe: A Case Study in Social Conflict and the State Policy in California Agriculture." *Social Problems* 30(1): 26–39.

Navarro, Vicente. 1982. "The Labor Process and Health: A Historical Materialist Interpretation." *International Journal of Health Services* 12(1): 5–29.

– 1978. *Class Struggle, the State and Medicine*. London: Martin Robertson.

Neff, James Alan, and Baquer A. Husaini. 1982. "Life Events, Drinking Patterns and Depressive Symptomatology." *Journal of Studies of Alcohol* 43: 301–18.

Neis, Barbara (Newfoundland Fishery Research Group). 1986. *The Social Impact of Current and Future Technological Change in Harvesting and Processing*. St. John's, Nfld.: Institute of Social and Economic Research.

Neutal, C. Ineke. 1989. "Mortality in Commercial Fishermen of Atlantic Canada." *Canadian Journal of Public Health* 80: 375–9.

Nicholas, T., and P. Armstrong. 1973. *Safety or Profit: Industrial Accidents and Conventional Wisdom*. Bristol: Falling Press.

Nie, Norman H., C. Hadlai Hall, Jean Jenkins, K. Steinbrenner, and D. Bent. 1975. *SPSS Statistical Package for the Social Sciences*. New York: McGraw-Hill.

Nolan, Bryon. 1973. "A Possible Perspective on Deprivations." In *Seafarer and Community*, edited by Peter Fricke, 85–96. London: Croom Helm.

Norr, James L., and Kathaleen L. Norr. 1978. "Work Organization in Modern Fishing." *Human Organization* 37(2): 163–71.

– 1977. "Societal Complexity of Production Techniques: Another Look at Udy's Data on the Structure of Work Organization." *American Journal of Sociology* 82: 845–53.

Norr, Kathaleen L., and James L. Norr. 1974. "Environmental and Technical Factors Influencing Power in Work Organizations." *Sociology of Work and Occupations* 1(2): 219–51.

Northwest Atlantic Fisheries Organization. 1989. *Scientific Council Summary Document* No. N1699. Dartmouth: NAFO.

Nova Scotia. *Trade Union Act*, S.N.S. 1972, c. 19.

Nova Scotia. *Workmen's Compensation Act*, R.S.N.S. 1967, c. 343, am. S.N.S. 1968, c. 5, now *Workers' Compensation Act*, R.S.N.S. 1989, c. 508.

Nova Scotia. Commission Regarding Workmen's Compensation Inquiry (Part III). 1968. *Report.* Truro, N.S.: The Commission. Mimeograph.

Nova Scotia. Committee on Occupational Health and Safety. 1984. *Report to the Minister of Labour and Manpower of the Committee on Occupational Health and Safety.* Halifax: The Committee.

Nova Scotia. Department of Fisheries. 1979. *Development of Nova Scotia's Fishing Fleet.* Halifax: Nova Scotia Communications and Information Centre.

Nova Scotia. Department of Labour. 1978–90. *Annual Report of the Workers' (Workmen's) Compensation Board of Nova Scotia.* Halifax: Department of Labour.

Nova Scotia. Royal Commission on Ratings of the Lunenburg Fishing Fleet and the Lumber Industry. 1927. *Report and Findings.* Halifax: Minister of Public Works and Mines.

Nova Scotia. Royal Commission on Workmen's Compensation. 1937. *Report.* Halifax: King's Printer.

Nova Scotia. Royal Commission to Inquire into the Workmen's Compensation Act of Nova Scotia. 1958. *Report.* Halifax: The Commission. Mimeograph.

Oakley, Anne. 1974. *The Sociology of Housework.* Bath: Martin Robinson.

Pfeffer, Jeffrey. 1981. *Power in Organizations.* Boston: Pitman.

Pilcher, William. 1972. *The Portland Longshoremen: A Dispersed Urban Community.* New York: Holt Rinehart and Winston.

Piven, Frances Fox, and Richard A. Cloward. 1971. *Regulating the Poor: The Functions of Public Welfare.* New York: Pantheon Books.

Poggie, John J. Jr. 1980. "Ritual Adaptation to Risk and Technological Development in Ocean Fisheries: Extrapolations from New England." *Anthropological Quarterly* 53(1): 122–9.

Poggie, John J. Jr., and Carl Gersuny. 1972. "Risk and Ritual: An Interpretation of Fishermen's Folklore in a New England Community." *Journal of American Folklore* 85: 66–72.

Poggie, John J. Jr., and Richard Pollnac. 1978. *Social Desirability of Work and Management among Fishermen in Two New England Ports.* Anthropological Working Paper no. 5. Kingston, R.I.: International Center for Marine Resource Development.

Pollnac, Richard. 1988. "Social and Cultural Characteristics of Fishing People." *Marine Behaviour and Physiology* 14: 23–39.

Pollnac, Richard, and John J. Poggie Jr. 1990. "Social and Cultural Factors Influencing Fishermen's Awareness of Safety Problems." in *The Proceedings of the First International Symposium on Safety and Working Conditions aboard Fishing Vessels*, edited by J.P. Roger, 407–12. Rimouski, Quebec: University of Quebec at Rimouski.

– 1988a. "Danger and Ritual Avoidance among New England Fishermen." *MAST* 1(1): 66–78.

– 1988b. "The Structure of Job Satisfaction among New England Fishermen and Its Application to Fisheries Management Policy." *American Anthropologist* 90(4): 888–901.

– 1979. *The Structure of Job Satisfaction among New England Fishermen.* Anthropological Working Paper no. 31. Kingston, R. I.: International Center for Marine Resource Development.

Popoff, T., and S. Truscott. 1986. *The Emotional Well-Being of Canadian Military Families in Relation to the Canadian Population.* Operational Research and Analysis Establishment Project Report PR364. Ottawa: Department of National Defence.

Popoff, T., S. Truscott, and R. Hysert. 1986. *Military Family Study: An Overview of Life/Work Stress and Its Relationship to Health and Organizational Morale.* Operational Research and Analysis Establishment Project Report PR351. Ottawa: Department of National Defence.

Proskie, John, and Janet Adams. 1971. *Survey of the Labour Force in the Offshore Fishing Fleet, Atlantic Coast.* Ottawa: Department of the Environment, Economics Branch, Fisheries Service.

Pugh, Derek Salman, ed. 1971. *Organization Theory.* London: Penguin Books.

Reasons, Charles E., Lois L. Ross, and Craig Paterson. 1981. *Assault on the Worker: Occupational Health and Safety in Canada.* Toronto: Butterworths.

Reimann, B., and G. Inzerilli. 1979. "A Comparative Analysis of Empirical Research on Technology and Structure." *Journal of Management* 5(1): 167–80.

Reschenthaler, G.B. 1979. *Occupational Health and Safety in Canada: The Economics and Three Case Studies.* Montreal: Institute for Research on Public Policy.

Riordan, Catherine A., G. David Johnson, and J. Stephen Thomas. 1991. *Handbook of Job Stress,* edited by P.L. Perrewe. *Journal of Social Behaviour and Personality* 6(7) (Special Issue): 391–409.

Rix, K.J.B., D. Hunter, and P.C. Colley. 1982. "Incidence of Treated Alcoholism in North-east Scotland, Orkney and Shetland Fishermen, 1966–70." *British Journal of Industrial Medicine* 39: 11–17.

Rodahl, Kare, and Zdenek Vokac. 1977a. "Work Stress in Long Line Bank Fishing." *Scandinavian Journal of Work, Environment and Health* 3: 154–9.

– 1977b. "Work Stress in Norwegian Trawler Fishermen." *Ergonomics* 20: 633–42.

Rodahl, Kare, Zdenek Vokac, P. Fugelli, O. Vaage, and S. Maehlum. 1974. "Circulatory Strain, Estimated Energy Output and Catecholamine Excretion in Norwegian Coastal Fishermen." *Ergonomics* 17(5): 585–602.

Rosenbaum, Paul D., and Judith Bursten. 1988. *Special Study on Labour Force Groups*. Canada's Health Promotion Survey: Technical Report Series. Ottawa: Health and Welfare Canada.

Rowbottom, B., and D. Billis. 1977. "Stratification of Work and Organization Design." *Human Relations* 30(1): 53–76.

Rubin, Lillian. 1976. *Worlds of Pain: Life in the Working Class Family.* New York: Basic Books.

Rummel, R.J. 1967. "Understanding Factor Analysis." *Conflict Resolution* 11: 444–80.

Salaman, Graeme. 1979. *Work Organization: Resistance and Control.* London: Longman.

Sass, Robert. 1979. "The Underdevelopment of Occupational Health and Safety in Canada." In *Ecology versus Politics*, edited by William Leiss, 72–96. Toronto: University of Toronto Press.

Schatzkin, Arthur. 1978. "Health and Labour Power: A Theoretical Investigation." *International Journal of Health Services* 8(2): 213–34.

Schilling, R.S.F. 1971. "Hazards of Deep-Sea Fishing." *British Journal of Industrial Medicine* 28: 27–35.

– 1966. "Trawler Fishing: An Extreme Occupation." *Proceedings of the Royal Society* 59: 405–10.

Silverman, David. 1970. *The Theory of Organizations: A Sociological Framework.* London: Heinemann Educational Press.

Silverman, David, and Jill Jones. 1976. *Organisational Work: The Language of Grading/The Grading of Language.* London: MacMillan.

Sinclair, Peter. 1988. *A Question of Survival: The Fisheries and Newfoundland Society.* St. John's, Nfld.: Institute of Social and Economic Research.

Smith, Cortland L. 1981. "Satisfaction Bonus from Salmon Fishing: Implication for Management." *Land Economics* 57: 181–96.

Smith, Dorothy. 1975. "The Statistics on Mental Illness: What They Will Not Tell Us." In *Women Look at Psychiatry*, edited by Dorothy Smith and Sara J. David, 73–120. Vancouver: Press Gang Publishers.

Smith, M. Estelie. 1988. "Fisheries Risk in the Modern Context." *MAST* 1(1): 29–48.

Stiles, Geoffrey. 1971. "Labour Recruitment and the Family Crew in Newfoundland." In *North Atlantic Maritime Cultures: Anthropological Essays on Changing Adaptations*, edited by R. Andersen, 189–208. The Hague: Mouton Press.

Suschnigg, Peter T. 1988. "Unionization, Management/Labour Relations and Lost-Time Accidents." *Occupational Health in Ontario* 9(2): 96–116.

Swartz, David. 1977. "The Politics of Reform: Conflict and Accommodation in Canadian Health Policy." In *The Canadian State*, edited by Leo Panitch, 311–34. Toronto: University of Toronto Press.

Taylor, Howard F. 1970. *Balance in Small Groups.* New York: Van Nostrand Rheinhold.

Thiessen, Victor, and Anthony Davis. 1988. "Recruitment to Small Boat Fishing and Public Policy in the Atlantic Canadian Fisheries." *Canadian Review of Sociology and Anthropology* 25(4): 601–25.

Thompson, Paul. 1985. "Women in the Fishing: The Roots of Power between the Sexes." *Journal for the Comparative Study of Society and History* 27: 3–32.

Thompson, Paul, Tony Wailey, and Trevor Lummis. 1983. *Living the Fishing.* London: Routledge and Kegan Paul.

Trice, Harrison M., and William J. Sonnestuhl. 1988. "Drinking Behavior and Risk Factors Related to the Work Place: Implication for Research and Prevention." *Journal of Applied Behavioral Science* 24(4): 327–46.

Truscott, S., and S. Flemming. 1986. *Occupational Stress among Married and Single-Parent Canadian Forces Personnel.* Operational Research and Analysis Establishment Project Report PR375. Ottawa: Department of National Defence.

Tunstall, Jeremy. 1962. *The Fishermen.* London: MacGibbon and Kee.

United Nations. Food and Agricultural Organization. 1987. *World Nominal Catches.* Vol. 64. Rome: FAO.

Walker, Lenore E. 1984. *The Battered Woman Syndrome.* New York: Spring Publishing.

Walters, Vivian. 1985. "The Politics of Occupational Health and Safety: Interviews with Workers' Health and Safety Representatives and Company Doctors." *Canadian Review of Sociology and Anthropology* 22(1): 57–79.

– 1983. "Occupational Health and Safety Legislation in Ontario: An Analysis of Its Origins and Content." *Canadian Review of Sociology and Anthropology* 20(4): 413–34.

– 1982. "Company Doctors' Perceptions of and Responses to Conflicting Pressures from Labour and Management." *Social Problems* 30(1): 1–12.

Warner, William. 1983. *Distant Water: the Fate of the North Atlantic Fisherman.* Boston: Little, Brown and Co.

Wilson, Jean D., Eugene Braunwald, Kurt J. Isselbacher, Robert G. Defersdorf, Joseph B. Maetin, Anthony S. Farci, and Richard K. Root, eds. 1991. *Harrison's Principles of Internal Medicine.* 12th ed. Vol. 1 and 2. New York: McGraw-Hill.

Winsor, Fred. 1987. *A History of Occupational Health and Safety in Nova Scotia's Offshore Fishery, 1915–1985.* MA thesis, St. Mary's University, Halifax.

Wood, Stephen, 1975. *The Degradation of Work? Skill, Deskilling and the Labour Process.* London: Hutchinson.

Zimbalist, Andrew, ed. 1979. *Case Studies on the Labour Process.* New York: Monthly Review Press.

Zwickers, Re; Re Application of Lunenburg Sea Products Ltd. (1947), 21 M.P.R. 305 (N.S.T.D.).

Index